Tomorrow's math

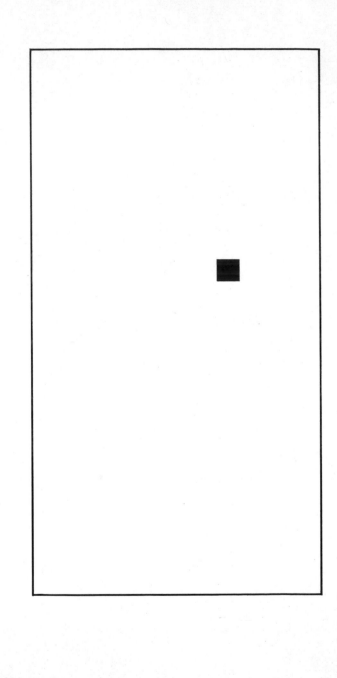

C. Stanley Ogilvy

Tomorrow's

SECOND EDITION

unsolved problems
for the amateur

New York
OXFORD UNIVERSITY PRESS
London   1972   Toronto

Preface to the second edition

Several problems from the first edition of this book have been solved during the decade since that edition appeared. These have been removed and in most cases the solutions are mentioned in the Notes. A few others have been moved out to make room for better problems; and many new ones have been added. I have again been impressed by the large number of unsolved problems, accessible to description in not very technical language, that are at large in the mathematical community today. There was an oversupply to choose from. I hope that you will be pleased with the new selection of these stubborn, sometimes strange, and always tantalizing creations.

C.S.O.

Clinton, N.Y.
July 1971

Contents

Tomorrow's math

1

The meaning
of an unsolved problem

What is a mathematical problem?

To a mathematician, a problem is often much more than "a question requiring an answer." He may refer to his whole current research project as a problem. If he does, it usually means that he is working on a "good" problem, one which contains many ramifications and subtleties and perhaps ties in with other problems. It may be a mistake to speak of good mathematical problems as if there were also bad ones; but there are certainly interesting problems and less interesting ones. The most attractive problems are usually relatively easy to state, have wide application either in mathematics or elsewhere, and possess a certain intrinsic charm of their own.

To say that a working mathematician spends all his time tackling unsolved problems would be a misrepresentation unless the idea of a problem is taken in the very broadest sense as *any* mathematical topic. Some mathematicians become so engrossed in one or more specialties that they never bother with separate problems, considering them a waste of time. Others like to look about a bit and are not above trying their hand at any poser which may have stumped a colleague. The fact remains that unsolved problems are a considerable source of inspiration in the field. Although we certainly cannot claim that all or even most of the problems presented in this non-technical book are of such high caliber, we shall still hope to indicate the kind of lure that some problems hold for the technician, while at the same time offering a sampling of material accessible to the amateur.

☐

In elementary arithmetic books one sees lists of exercises sometimes headed *Problems:* "A rectangle measures 5 inches by 7 inches. What is its area?" The "answer" to this "problem" is 35 square inches. A child might obtain the answer by carefully drawing a 5-inch by 7-inch rectangle, ruling it off into square inches, and counting the squares. Thus stated and thus solved, the problem would not satisfy a mathematician, principally because the solution would be of no help in finding the area of another rectangle, say one 6 inches by 10 inches. Thus to the mathematician a solution is usually not an *answer* but a *method*. A better problem is, given the lengths of the sides of a rectangle, how does one find its area?

A still better problem is, does every plane rectangle have associated with it a numerical quantity which can meaningfully be called an area? Note that the previous problem cannot be solved unless the answer to this one is yes. Yet very little emphasis is put upon this last question in elementary mathematics. Its answer seems obvious: it is taken for granted.

Since about the middle of the last century, mathematicians have become extremely wary of taking anything at all for granted. Accepting something as "obviously true" has led them astray too often. For instance, it is certainly obvious that every piece of paper, like this page, has two sides, in the sense that a bug crawling on one side could not get to the other side without passing around an edge or boring a hole through the paper. Obvious—but false. Perhaps you have heard of a Möbius band. Take a long narrow ribbon of paper, bring the ends together, and join (glue) them to form a ring (Figure 1). If A is brought into coincidence with B, and C with D, then an ordinary band results, with an inside surface and an

[1]

outside surface. But before gluing, give the strip a half-twist, so that A is joined at D and B at C. The resulting surface is a Möbius band: it has only one side and one edge, and a bug could crawl all over it without ever crossing that edge. An interesting (and easy) problem is, does such a surface have an area, and if so, how is it found?

That this matter of area is by no means trivial becomes clearer when we get into curved surfaces. Does a sphere have a surface area? If so, how can it be defined? If it can be defined, how can it be measured? At least partially as a consequence of this question, an extensive non-elementary branch of mathematics called measure theory has been developed. You may think that you know intuitively that a curved surface (the earth's, for example) has an area. But even if the surface in question is smooth, like that of a mathematical sphere, how can you measure it? It cannot be flattened out and hence measured by comparison with a known plane area. The mathematical expression is that a spherical surface is not *developable* from a plane. You may say, "I could paint the surface, measuring carefully the amount of paint used"; but paint has thickness, and by applying a *shell* to the surface you would be destroying the concept of area. To be sure, one has a formula* for the area of a sphere in terms of its radius; but,

---

* $S = 4\pi r^2$. Archimedes considered it his best discovery.

what confidence can be placed in such a formula until the concept of surface area has been satisfactorily defined?

You probably think that I am only teasing, and that mathematicians have surely succeeded in defining the area of any curved surface by this time. Strangely enough, that is not the case: a simple yet adequate mathematical definition satisfactory for all purposes has yet to be devised. We pursue the matter further in the notes.

Other difficulties arise in connection with volumes of solids. Intuitively it would seem that if you start with a given finite solid object, cut it up into a finite number of pieces, and then reassemble these same pieces in any way, provided there are no spaces anywhere between the reassembled pieces the new solid should have the same volume as the original. In 1924 Banach and Tarski showed that this is not necessarily so by proving the following extraordinary theorem: it is possible to cut a solid sphere into a finite number of pieces and reassemble them by rigid motions (no distortion) to form two solid spheres (no holes), each of the same size as the original one. It might be supposed that the number of pieces in the dissection would have to be very large. But Raphael Robinson has shown that there need be no more than five pieces! These unbelievable results show that we must revise our fundamental notion of volume. There cannot be any completely general definition which will preserve volume under rigid motion—something always previously accepted as "obvious."

Here is a much simpler illustration of the necessity of being on guard against the obvious. Of the five regular "Platonic" solids, one has 12 (pentagonal) faces and another has 20 (triangular) faces. Because of their regularity, both can be inscribed in a sphere so that every vertex touches the surface of the sphere. Suppose we inscribe each of these two solids in the same sized sphere; which solid will have the greater volume? Each one of the faces of the 20-hedron must be much smaller than each one of the faces of the 12-hedron. That happens to be true. The more numerous smaller faces must "cling more snugly" to the surface of the sphere than the fewer larger faces; and hence the volume of the 20-hedron, being nearer to that of the sphere, must be greater than the volume of the 12-hedron. That happens to be false. The volume of the 12-faced solid is nearly 10 per cent larger than that of the 20-faced solid inscribable in the same sphere. Most people make the opposite guess.

□

Perhaps you are beginning to agree that some "intuitively evident" facts turn out to be not facts at all and may lead to really difficult problems. Euclid's postulates for plane geometry were originally supposed to be "self-evident truths." Today's mathematician no longer has anything to do with "truths" as universals; they are left to the philosophers. But Euclid's postulates were supposed to be true in the sense that any straight-thinking human being would accept them as obvious ("a straight line is the shortest path between two points," and the like). The fifth postulate, however, dealing with parallels, has a somewhat different flavor. It states, in effect, that through a point not on a straight line one and only one line can be constructed parallel to the given line. People sensed that this postulate should be *deducible* from the others, that is, could be proved as a theorem. For centuries many mathematicians attempted to solve this challenging problem. The ultimate solution is a negative one: the parallel postulate is indeed a postulate on its own merit, and cannot be deduced from the others. This conclusion was not reached until it was realized, about a hundred years ago, that the problem was being attacked from the wrong direction. If it is only a postulate, one should be able to replace it by some *other* postulate and develop an equally sound geometric system. This can in fact be done. It was unwillingness to face—indeed, inability to conceive—the possibility that there might exist other geometries than Euclid's that clouded the issue for two thousand years. Once this hurdle had been surmounted, progress was rapid.

If an unsolved problem has been on the

books for a long time, it is very often difficult because it is being presented in a way in which it can never be answered. The stroke of genius that creates an entirely new approach to an old familiar problem is exceedingly rare. Sometimes, after such a new attack solves a problem, it seems easy and we wonder why no one thought of it before. Usually the trouble was that we had been looking in the wrong direction; often we had been seeking entirely the wrong answer.

One of the three famous problems of antiquity was that of the trisection of an angle with ruler and compass. There are many known methods of trisecting an angle, but the problem was how to do it with only the classical tools, compass and straightedge. For centuries mathematicians sought the wrong answer; for it cannot be done. Even when this was suspected, the search could not proceed in the right direction until the correct methods were devised. We shall return to this problem.

☐

Success in mathematics is the result of clever ideas. Breakthroughs are made sometimes by means of sheer plodding persistence, but more often, as in all sciences, through a flash of brilliance. Such flashes, however, do not happen by themselves. They are the result of hard concentration, backed by years of training in mathematical thinking. What does that mean?

The meaning of an unsolved problem

Just how does a mathematician think? If he is a good one, he is ever on the alert for a new way of looking at problems, what L. A. Graham has called "the surprise attack." We attempt to illustrate this with two easy examples.

Suppose that there are 64 players entered in a tennis tournament that is to be set up in the usual way. For the first round the players are paired by lot to play 32 matches. The winners of this round meet to play the 16 matches of the second round, and so on, to the finals. The total number of matches played is thus $32 + 16 + 8 + 4 + 2 + 1 = 63$. This can be rewritten

$$2^0 + 2^1 + 2^2 + 2^3 + 2^4 + 2^5 = 2^6 - 1,$$

and the rule is perfectly general. 64 is a power of 2; and whenever the number of players is exactly a power of 2, the total number of matches is that power of 2, minus 1.

But what if some other number of players enters the tournament, 47 or 93 or whatever? A certain number must now sit out the first round; that is, they "draw a bye" into the second round, which is arranged to come out a power of 2, and from then on the tournament runs as before. The problem is, what is the total number of matches that have to be played? What is a *general* formula for the number of matches, when the number of entries is not a power of 2?

This can hardly be classed as a respectable

mathematical problem; it is too easy. By trying various examples, and counting, you could doubtless arrive at a conclusion. Or you could calculate the number of byes required, as a function of the number of entries and the required power of 2 for the second round, and thus derive a formula.

The point is that the problem has also an inspired solution, a mathematician's solution, that requires no computation, no formulas, no numbers—just pure thought. How many *losers* does each match have? Exactly one. How many times does *each player* in the tournament lose a match? Exactly once—except the tournament winner, who never loses. Therefore the total number of matches, which equals the total number of losers, is always one less than the number of players. An interesting minor feature of this solution is that it is independent of the method of structuring the rounds. Therefore it proves the additional fact that even if the officials set up the tournament incorrectly the number of matches is unchanged. Suppose there are 40 players. An inexperienced director might arrange all the players in 20 first-round matches, bye some of the 20 winners from the second round into the third, and then do it right from there on. No matter where the byes are allowed to take place, the total number of matches will always be 39.

The second example is from elementary geometry.

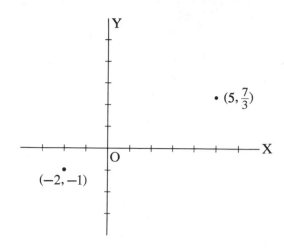

[2]

Each point of the (x, y) plane is designated by its coordinates: $(5, \frac{7}{3})$ means the point that has $x = 5$, $y = \frac{7}{3}$ (Figure 2). All those points both of whose coordinates are integers are called lattice points. Thus $(-2, -1)$ is a lattice point but not $(5, \frac{7}{3})$. Suppose now we consider the orthogonal projection of a point upon the 45° line. That is, if we drop a perpendicular from point P (Figure 3) to the line $y = x$, so that OQP is a right angle, then Q is the projec-

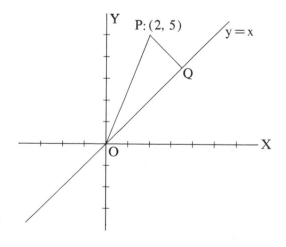

[3]

tion of P. The theorem is this: If P is a lattice point, then Q always has coordinates $(n/2, n/2)$, where n is an integer. Of course the two members of the pair of coordinates of any point on the line $y = x$ are equal to each other; what we are claiming is that all points Q are either lattice points themselves or are of the form $(n/2, n/2)$ with n an odd integer.

Unless you happen to think of the right approach this may not be self-evident. It can be proved "the hard way" by considering the right triangle OQP, and making calculations based on the hypotenuse length and the given $45°$ angle between OQ and the x-axis. Eventually this will drive out the result. The mathematician's way, however, might be the following. If P is a lattice point, say $(2, 5)$, then P', its reflection in the line OQ, is *also* a lattice point, in this case $(5, 2)$. (See Figure 4.) But Q is the midpoint of PP'; and the well-known midpoint formula says that its x-coordinate is the average of the x-coordinates of the end

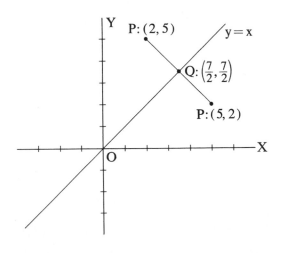

points, and similarly for the y-coordinates:

$$x = \frac{x_1 + x_2}{2}, \qquad y = \frac{y_1 + y_2}{2}.$$

In this case we have

$$\left(\frac{2 + 5}{2}, \frac{5 + 2}{2}\right) = \left(\frac{7}{2}, \frac{7}{2}\right),$$

as claimed. Whenever the numerator is even we have lattice points, when it is odd we have points $(n/2, n/2)$.

A slightly different "quickie" solution is suggested by the diagram in the Notes.

□

One of the most successful general methods for solving a troublesome problem is that of *embedding* it in a more general context than that in which it first appeared. To cite a simple example, an arithmetic problem may become more comprehensible when treated by the more powerful methods of algebra. The numbers 98 and 102 can be multiplied together by the usual procedure, and the answer somewhat laboriously obtained; but by fitting the exercise to an algebraic form, one can do it mentally. It is only necessary to recognize $(100 - 2)(100 + 2)$ as a special case of $(a - b)(a + b) = a^2 - b^2$ to read off the answer $10000 - 4 = 9996$.

A more impressive example of embedding

arose historically in connection with convergence of infinite series. If we divide, by ordinary algebraic long division, we get

$$\frac{1}{1-x} = 1 + x + x^2 + x^3 + \cdots$$

where the dots mean "and so on." To check that this *might* have meaning in spite of the non-terminating quotient, try multiplying back:

$$
\begin{array}{l}
1 + x + x^2 + x^3 + x^4 + \cdots \\
1 - x \\
\hline
1 + x + x^2 + x^3 + x^4 + \cdots \\
\quad - x - x^2 - x^3 - x^4 - \cdots \\
\hline
1 + 0 + 0 \; + 0 \; + 0 + \cdots
\end{array}
$$

So it appears to check. However, we must look further at the quotient series. For what values of x, that is to say what size x's, does the series converge? It is a geometric progression, and you may recall from your high school days that an infinite geometric progression converges if and only if the common ratio of two successive terms is numerically less than 1. Here the ratio is x. If we multiply any term by x we get the next term; hence the series converges, and in fact the limit of the sum is $\frac{1}{1-x}$, whenever the numerical value of x is less than 1, written $|x| < 1$. We would not expect a valid equality for $x = 1$ for two reasons: the series $1 + 1 + 1 + 1 + \cdots$ obviously *diverges* (does not approach any fixed number); and the expression on the left becomes $\frac{1}{0}$, which has no meaning; we are not allowed to divide by zero. The fraction $\frac{1}{1-x}$ does have meaning for $|x| > 1$; but it no longer represents the infinite series $1 + x + x^2 + x^3 + \cdots$ for such values. An adequate warning that things were going bad was given when $\frac{1}{1-x}$ "misbehaved" at $x = 1$.

Now, again by long division, suppose we obtain

$$\frac{1}{1+x^2} = 1 - x^2 + x^4 - x^6 + \cdots$$

Once more the right-hand side is a geometric progression, and again it converges only when $|x| < 1$. But this time the left-hand side gives *no indication* that anything might go wrong at $x = 1$. If this value is substituted into the series, it fails to converge; yet the left-hand side becomes simply ½. It was not clear to mathematicians what was going on here until the number system was extended. When the real numbers were embedded in the complex domain then the difficulty disappeared. The denominator $1 + x^2$ equals zero when $x^2 = -1$, or $x = \sqrt{-1}$. In the complex domain, the absolute value of $\sqrt{-1}$ is 1. Thus we have a value of x such that $|x| = 1$ which makes the left-hand side "go bad," just as in the first example, and the warning has been issued not to take $x \geqslant 1$.

We return briefly to the problem of the trisection of an angle. An outline of the proof that such a construction is impossible goes as follows. First one shows that ruler and compass constructions are limited in their scope, in that they can produce magnitudes expressible with radicals *only* if the expression contains a finite number of square roots and combinations thereof but no other kinds of roots. Now if we are given an angle θ to trisect, we must find $\theta/3$. By trigonometric identities, one obtains $\cos \theta = 4 \cos^3 \theta/3 - 3 \cos \theta/3$. Hence if $\cos \theta$ is given, say k, the problem is equivalent to solving the equation $4x^3 - 3x - k = 0$, a cubic. This equation is irreducible; therefore its solutions are *not* expressible in terms of a finite number of square roots, and hence cannot be constructed with ruler and compass.

Mathematicians were unable to reach this conclusion by purely geometric considerations. It was by embedding the question in an analytic medium that the result was obtained. To show that a problem cannot be solved is in itself a solution. Thus the classical trisection is impossible, and the problem has been disposed of once and for all.

Some constructions are impossible in a different sense. For example, suppose we are given a straight line with two points A and B

marked on it. What is the shortest path that starts at *A*, perpendicular to *AB*, and ends at *B*? There is no shortest path; although the *lower limit* of all possible paths has length *AB,* this lower limit cannot be attained without violating the perpendicularity requirement. Whatever path is chosen there is always another shorter one (see Figure 5). In this example, the non-existence of the sought solution is itself the ultimate solution.

A less trivial example of this kind is worth mentioning. What is the closed curve of minimum length passing through the midpoint *M* of the base of a given rectangle and dividing the area *A* of the rectangle into two simply connected parts? It is to be understood that the curve may touch the edges of the rectangle but must not wander outside its boundaries. The altitude from *M* will not do because it is not a closed curve; yet it seems that the problem ought to be solvable.

Figure 6a shows a closed curve through *M* dividing *A* into two parts such that the shaded

[5]

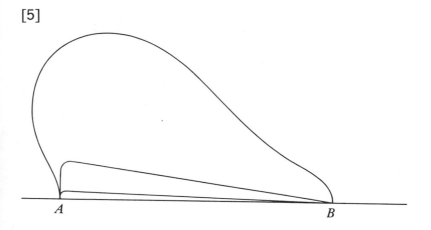

A   *B*

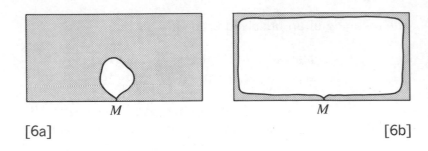

M

M

[6a]   [6b]

area is certainly larger than the white area; in Figure 6b the shaded area is the smaller of the two. There are many curves which make the two areas equal. Of them, which one is the shortest? It turns out that the best curve is the circle through M, tangent to the base, with area equal to $\frac{1}{2}A$. But this means that if the ratio of length to breadth of the rectangle exceeds $\pi/2$, the problem has no solution: the required circle will not fit inside the rectangle, some other curve must be used, and there is no shortest one.

Forty years ago Kurt Gödel startled the mathematical world by producing a new kind of unsolved problem. It had always been taken for granted (again that dangerous assumption!) that all mathematical propositions could be proven to be either true or false. What Gödel did, in effect, was to show that there exist mathematical propositions which make perfectly good sense and which may be true or may be false (in the mathematical sense of compatible with or not compatible with a set of stated assumptions), but for which there is no way of ever finding out whether they are true or false. A problem of this kind is *undecidable*, in that one cannot say whether it has a solution or not. Many mathematicians believe that problems unsolvable in the Gödel sense are not of frequent occurrence nor easily stated. It seems possible that none of the problems in this book will turn out to be Gödel-undecidable.

☐

Our order of presentation is somewhat random. We shall mention some problems that can be stated and understood with a minimum of mathematical background, and move from them toward problems that cannot be adequately explained without some previous mathematics, which we shall attempt to supply as needed, at least in skeletal form.

2

Applied problems

We begin with a few problems that might be labeled "practical." Just how practical they are depends on the point of view; at least they can be related in some direct way to everyday life, and hence if we wish we can categorize them as among the problems of applied mathematics.

Formerly a country stationed its defensive troops in strategic positions along its border. Today, since attack has become three dimensional, one has to anticipate bombings and paratroop landings within the interior of the country. The question has therefore become important: How should one's defenses be distributed?

To simplify the problem, suppose a country with a large land area and a small population anticipates an air invasion; planes can land anywhere; there is no one spot more likely to be selected by the enemy than any other. The home army has a limited number of mobile defense units. How should they be deployed around the countryside? The enemy will find out where they are located and will land at the point whose distance from the nearest station is greatest. Therefore the object of the defenders is to place stations in that pattern which renders the greatest distance from any one point to some station a minimum. We shall call this the pattern of *best deployment*.

The problem is far from completely solved. Even the answer to this simpler question is unknown: What is the best deployment of n stations (points) on a plane circular disk? More precisely, no point of the disk is at a distance greater than k from some one of the stations; what is the smallest possible value of k for various n? The answer is known for $n \leqslant 5$, but there are many values of $n > 5$ for which the smallest k is not known, and a general solution seems remote at present.

On a sphere the problem becomes somewhat more interesting if we ask instead for the best *dispersal* of n points; that is, the arrangement that makes the least distance between any two *of them* a maximum. The deployment problem asks how best to guard the area with stations; the dispersal problem asks how best to keep the stations apart, like (perhaps) an equilibrium pattern for mutually repellent particles. This may sound like two ways of saying the same thing, but it is not. One might guess that n points could always be dispersed on the sphere more sparsely than $n + 1$ points, but the guess is incorrect. It is not possible to scatter five points so that every pair of them is separated by more than 90° of arc. In other words, 90° is the optimum mutual separation. Yet the same statement is true of six points, which highlights the difference between dispersal and deployment. Even on a sphere, which could be considered a country with no border, the value of k in the deployment problem is not the same for $n = 5$ as it is for $n = 6$.

The values of n for which the dispersal problem on the sphere has been solved are 2 through 12, and 24. Solutions for other values of n, or some form of general solution, are unknown. Interestingly, the solution (recently discovered) for $n = 11$ is that there is no improvement over $n = 12$; the situation parallels the 5, 6 case.

These problems, previously thought to be of only theoretical interest, have recently acquired new practical aspects. (1) If one wishes to monitor outer space, either from one country or from the whole globe, what is the best arrangement of the listening posts (radio telescopes)? (2) The communications industry is interested in placing a large number of echo satellites into orbit for the purpose of reflecting or retransmitting line-of-sight radio waves. If these are randomly scattered, what will be their expected average separation at any instant? Is there any optimum arrangement of the orbits? (3) In the event that, some day, the reflector satellites could be replaced by space stations under their own power, which could each be held over a specified spot on the surface of the earth, what would then be the best network of such stations?

A related problem troubled Isaac Newton, who guessed the right answer but was unable to prove it. How many spheres, solid physical balls if you wish, can be arranged to touch a single central sphere, all the spheres being of the same size? Note that if congruent spheres are tangent, each subtends on the other a circular cap of angular radius 30°. Hence the problem is equivalent to asking, how many non-overlapping caps of angular radius 30° can be placed on a sphere? One way of arranging such caps is to maximize the minimum distance between centers, and the optimum dispersal of 12 points leaves more than enough room for twelve 30° caps. It was not until two centuries after Newton's time, however, that it was proved that there was not enough room for thirteen caps.

An extension of Newton's problem is still open today. Replace the central sphere by two tangent spheres, and ask how many spheres can be placed tangent to the figure formed by these two spheres. The conjectured answer is 18. The two central spheres must be tangent to each other; each of the other spheres must be tangent to at least one of the central spheres; all the spheres must be of the same size; and there must be no overlapping.

□

How should three cowboys, who have to watch over cattle on a square range, station themselves? This is readily recognizable as the deployment problem for a square area with $n = 3$. Hugo Steinhaus has elaborated the problem by imposing additional intriguing restrictions. First he points out that dividing the square into three equal rectangles with a cowboy at the center of each rectangle (Figure 7) has four *advantages:* (1) The areas are equal. (2) The maximum rides (distance each cowboy has to go to reach the farthest point in his area) are equal. (3) Each point in the range is entrusted to the nearest cowboy. (4) Each

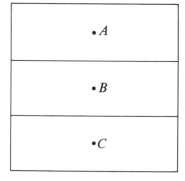

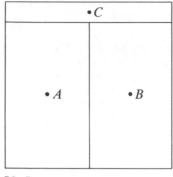

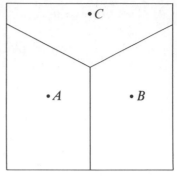

[8a]   [8b]

cowboy is stationed at the point in his area that minimizes his maximum ride. He next considers five other partitions of the range, each of which reduces the maximum ride but fails to have at least one of the four advantages. Finally he poses the question, of all those partitions possessing all four advantages, are there any with a shorter maximum ride than that of Figure 7?

It has been shown that the solution of the unadorned problem is $k = \sqrt{65}/16$ for the unit square. Figure 8a and b shows two possible partitions with this value of k. The first fails to meet advantage (3), and both violate advantage (1). This suggests that if the answer to the Steinhaus question is yes, then the maximum ride for any partition that answers it may turn out to be greater than $\sqrt{65}/16$.

☐

What course should a ship steer in order to meet a second ship as quickly as possible? The courses of both ships are to be straight, and the first ship (pursuer) is assumed to be faster than the second ship (quarry), but both are to maintain constant course and speed. The solution for the problem thus simplified is well known. Let P and Q be the positions of pursuer and quarry respectively at the instant when pursuit begins. If the quarry steers any straight course, the pursuer can intercept soonest by steering another straight course aimed at the point where the fugitive's course intersects an Apollonian circle associated with P and Q. This circle is the locus of points k times as far from P as from Q, where k is the ratio of the ships' velocities (Figure 9). Note that Q is not the center of the circle.

The above solution is applicable only to a plane ocean, and hence is practical for short interceptions. If the plane theory is adapted to the surface of a sphere, interception will not in general occur. To identify or characterize the locus of collision points on a sphere for constant speeds and great circle courses re-

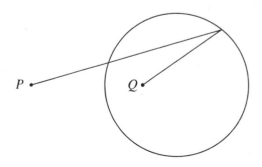

[9]

mains an unsolved problem. It may find application in long-range, high-speed air navigation where the earth's sphericity cannot be neglected.

☐

A man falls overboard from a boat in a thick fog, into a wide river with parallel banks which he cannot see. He knows the width of the river but not his distance from either shore. What is his best procedure? Under the somewhat unrealistic assumption that he is able to navigate—that is, to keep track of his course—what is the shortest course he can swim and be sure of reaching land?

Another problem is posed by the boat lost at sea at a known distance but unknown direction from the nearest shore, which is this time a single straight beach. This could happen to a careless fisherman who has run out to sea a known number of miles and stopped to fish; while he is busy netting a big one, the fog settles down, so that when he is ready to go home he realizes that he has completely lost his bearings. We have to assume that he is without a reliable compass, else he would not be truly lost; still, we must stipulate that he can somehow navigate, and we ask the same question that we asked in the swimmer's problem.

The conditions are altered in a quite different version: the boat has only a limited supply of gasoline, not enough to reach shore by the answer to the previous problem under worst possible luck, but ample to reach shore in a straight line if the boat happens to start out in the right or nearly the right direction. What now is the best plan? That is, what course should be pursued in order to maximize the probability of reaching land before the gas gives out?

The swimmer's problem could be altered in the same fashion by postulating that the man can swim only a fixed (known) distance before becoming exhausted.

□

While on the subject of boats we insert some questions of a very different nature drawn from the world of sport. They are part of what is known as the ranking problem.

The rules of the International Star Class Yacht Racing Association state that a series of races shall be scored by totaling the points of each boat in each race; that in each race a boat receives one point for finishing and one point for each boat defeated; and that a tie shall be resolved in favor of the boat that has beaten the other the greatest number of times. Assuming that there are no dead-heat finishes, find the conditions under which there can be an n-way, unresolvable tie in an m-boat series of r races.

By an *unresolvable* tie is meant a tie that remains even after the application of the above rule. Example: in a three-race series, A defeats B twice, B defeats C twice, C defeats A twice, and all three happen to amass the same total points. This is entirely possible, and in fact it has occurred several times in recent years in actual series of races.

(1) If $n = r$ there can always be an unresolvable tie, provided at least n boats are racing. Take the order of finish of any n boats in the first race; let each advance one position cyclically, so that the first goes to last place among the n; repeat this process for each race, and the result will be an n-way, unresolvable tie.

(2) If $r = 2$ there can always be up to an m-way, unresolvable tie: simply let the order of finish be reversed in the second race.

Both these examples illustrate *symmetric* ties, in which each tied boat beats each other tied boat the same number of times. But it is perfectly possible to have an *asymmetric* tie, provided there are enough other boats racing. Example (3) shows a three-way asymmetric tie in a five-race series of five or more boats. A beats B four times, B beats C three times, and C beats A three times. The tie is not resolved in favor of A because, under the rule, C has beaten A.

(3)

	Daily points					Total
A:	3	3	2	2	1	11
B:	2	2	1	1	5	11
C:	1	1	3	3	3	11

It appears that there must be at least five boats racing in order for B to make up the minimum deficit of four defeats, but even this is not proven. It may be possible to arrange it some other way, with only four boats. If there are seven races, four boats are certainly enough:

(4)

	Daily points							Total
A:	3	3	3	2	2	1	1	15
B:	2	2	2	1	1	4	3	15
C:	1	1	1	3	4	3	2	15

If r is composite (6, 9, etc.) the regatta can be decomposed into its prime factors for part of the answer. For instance if there are nine races there can be a three-way tie without auxiliary boats, because each of the three could repeat twice more the situation of example (1). In practice one seldom sails more than seven races in any series that is scored in this fashion, so that for higher values of r the interest is theoretical. The general theory has not been explored.

A somewhat similar problem is encountered

in making up a round-robin tournament schedule. J. E. Freund has presented a "simple way of constructing round robin schedules for any number of teams." The problem becomes more complicated, however, when an additional condition is imposed. Suppose a partnership is defined as a pair consisting of a man and a woman. Can two teams of eight partnerships play at eight bridge tables so that each person plays with new opponents *and* a new partner each time? Freund's treatment does not consider this version.

☐

A traveling salesman wishes to start from Washington, D.C., visit every state capital in turn, and return to Washington. How should he plan his intinerary to make the trip as short as possible? One might suggest programming the problem for a computer; then let the machine calculate the lengths of all possible trips and simply select the shortest. But for the continental United States alone there are 48! (meaning $48 \times 47 \times 46 \times \cdots \times 2 \times 1$) possible routes, too many for even a large computer to consider one by one. There is still no general way to solve the traveling salesman problem for an arbitrary number of cities on any map, although "a good many mathematicians have wrestled [with it] for more than twenty years."

☐

Applied problems

The design of an electrical switching circuit can be elegantly represented by a special kind of simple mathematical equation. The required Boolean algebra is explained in elementary language in the reference in the notes. The type of problem involved is illustrated by the following example. We wish to design a circuit connecting Terminal No. 1 with Terminal No. 2. Let A refer to all switches of type a, B to all switches of type b, and so on. Current is to flow if D is closed, provided also at least one of A, B, or C is closed; and also if D is open, provided *all* of A, B, and C are closed; but not otherwise.

One way to design the network is shown in Figure 10a; but the conditions are also fulfilled by the so-called bridge circuit of Figure 10b, with two fewer switches. As well as requiring less "hardware," a circuit without unnecessary switches has other important practical advantages. The problem of minimal, or most economical, network design leads to many unanswered questions. Some of these are: (1) Is there a general way of designing a switching circuit satisfying given requirements and employing a minimum number of switches? At present there is no known way, other than trial and

[10a]   [10b]

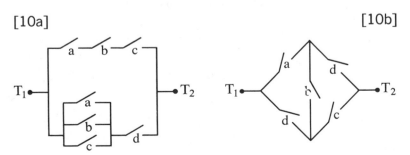

error, of knowing whether you have hit upon the circuit that cannot be further improved. (2) The same question using series-parallel systems only (no bridges). (3) The same question when there are certain on-off combinations which never actually occur in the use of the network. (4) Is there any way of predicting whether a minimal circuit for a given problem is going to be of bridge type or series-parallel or mixed?

□

An intriguing unanswered question is known as the paper bag problem. Is there any way to collapse a plane-faced linearly hinged container into 2-space (a plane), without deforming the faces? You may think that the paper bags as they come folded at the super-market solve the problem. But in order to unfold one of those bags the paper sections must be warped out of flat in places. That is, if the plane segments making up the bag could be bent strictly only at the creases, the folded bag would not unfold. Actually we add another condition that disqualifies the ordinary paper bag because it has an open top. Our container must be able to hold something (like water, for instance, if the paper could be made waterproof) even in a gravity-free environment. All facets of the object must remain planar at all times, which means that the linear hinging is actually a

redundant condition, because any two intersecting planes always meet in a straight line.
The problem is, is such a container theoretically possible? Amusingly enough, in the next
lower pair of dimensions, the solution is trivially simple. Any parallelogram is a closed
container for two-dimensional water, and it
collapses into 1-space.

□

Despite the best efforts of many of the greatest
mathematicians of the past two centuries, a
satisfactory theory of the motions of the moon
has eluded everyone. The intrinsic difficulty
of the problem has led to its virtual abandon-
ment, astronomers being content instead to
rely on numerical methods which can make
accurate predictions for a limited time. But
the problem has taken on new significance in
the calculation of orbits of man-launched or ar-
tificial satellites; and in this form, still unsolved,
it demands increased attention.

□

In the past, on college campuses where the
fraternity system flourished, it was customary
to allow each fraternity to choose more or
less freely whom it should invite to become
members. There were rules and "codes of
rushing," but even so some inequities were

bound to occur, and many undergraduates never had a chance to get into any fraternity. In an attempt to modernize the system, several colleges have recently instituted "total opportunity" schemes, whereby each student has at least the option of joining some fraternity if he wishes to. After the fraternities look over the candidates and vice versa, the fraternities submit lists of names and the students indicate their order of preference to a central committee. "Rushing" takes place according to a highly organized and rigid procedure designed to produce a maximum of suitable fits and a minimum of disappointing misfits. It is clear that some rather complicated mathematics is involved in designing an adequate system. Disregarding the difficulties due to human frailties and fickleness, not even the theoretical part of the problem is completely solved.

Almost the same problem is encountered by the college admissions offices in their attempts to select suitable freshman classes from among the eligible applicants. To date no central or national co-ordinating bureau of admissions exists, partly because no one has offered a convincing mathematical solution of the problem, although there are indications that a realistic solution is by no means unattainable.

3

Problems concerning games

Because of the mathematical structure of some games, interesting problems arise or can be invented around them. We look into a few of these, and then conclude the chapter with a brief mention of the new field of mathematics called Game Theory.

This chess problem was given to me by Martin Gardner, who says it is an old favorite but still unsolved. From the 16 black pieces, remove the 8 pawns, and place the remaining black pieces on the board in such a way that every space, including the occupied ones, is challenged by at least one piece. The two bishops must go on opposite colors, just as in actual play. It is not difficult to challenge 63 squares in this way; whether all 64 can be "covered" is not known.

□

The next problem is a paradox.

The headmaster of a school announced that some afternoon during the following week, on one of the weekdays Monday through Saturday, inclusive, there would be a surprise examination in a certain subject. The exact day of the exam was to be kept secret from the students, because if they knew which afternoon it was to be, they could do some last-minute cramming in the morning, and this the headmaster wished to prevent. He made it a sporting proposition by promising (and we are to assume that a headmaster's promise is binding), that if any student came to him during some morning and told him that the exam was to be given that day, and supplied him with a rational explanation leading to this conclusion, then he would cancel the exam, since it would no longer be a surprise.

The students got together over the weekend and came up with a brilliant idea. They nominated Algernon to present it to the headmaster on Monday morning.

"Sir," said Algy, "I am happy to inform you —I mean I am afraid you cannot give us that exam at all."

"No?"

"Will you agree that it would not be a surprise if you saved it until Saturday? Surely if we had not yet had the exam, we would know on Saturday morning that it must come that afternoon, since it would be the only day left.

One of us would come and tell you so, and you would have to cancel it."

"Very true. I cannot use Saturday, for the reason you give."

"So we may as well cross Saturday off the calendar? It is no longer an available day."

Reluctantly, seeing which way the argument was leading him, the headmaster had to agree.

"Very good," said Algy. "The week, effectively, starts on Monday and ends on Friday. But the very fact that you cannot use Saturday for the exam renders Friday useless also. For on Friday morning, not yet having had the examination, we would have to expect it on Friday afternoon. Thus Friday's exam could not be a surprise, and must be cancelled. So Friday must be crossed off the calendar too."

"You need say no more," said the headmaster, in deep perplexity. "I see that you can carry the argument back through the week, rendering each day unusable. It appears that no surprise exam is possible."

Algernon reported triumphantly to his colleagues that it had all worked out exactly as they had figured it must.

The paradox, however, is this. Although the above logic seems unassailable, what is to prevent the headmaster from selecting in advance some day at random, say Wednesday, which he reveals to no one? When Wednesday morning rolls around, how is any student to know that the exam will be given that afternoon?

Will not all conditions for a surprise have been fulfilled? No explanation of these apparently conflicting answers has been given, although more than one learned paper has been written in an attempt to resolve the paradox.

□

A Graeco-Latin square was originally an array made up of Greek letters and Roman letters. One can use any two distinguishable sets of n objects each, say letters and numbers, to form an n by n square. In each compartment of the square appears one letter and one number; but none appears twice in a single column or row. Furthermore, each number may be paired only once with each letter. Figure 11 shows a Graeco-Latin square of order 3.

Leonard Euler proved that such squares of odd order are always possible, and also of even orders divisible by 4. Graeco-Latin squares of even orders not divisible by 4 (6, 10, 14, etc.) he conjectured to be impossible. It is true that one cannot construct such a square of order 2

A 1	B 2	C 3
B 3	C 1	A 2
C 2	A 3	B 1

[11]

00	47	18	76	29	93	85	34	61	52
86	11	57	28	70	39	94	45	02	63
95	80	22	67	38	71	49	56	13	04
59	96	81	33	07	48	72	60	24	15
73	69	90	82	44	17	58	01	35	26
68	74	09	91	83	55	27	12	46	30
37	08	75	19	92	84	66	23	50	41
14	25	36	40	51	62	03	77	88	99
21	32	43	54	65	06	10	89	97	78
42	53	64	05	16	20	31	98	79	87

[12]

or 6; but in 1959 it was shown that all the others are theoretically possible; and one of order 10 was actually produced.

Instead of using a letter and a number in each compartment, we can use a first digit and a second digit, where now 4 7 does not mean forty-seven; it means the object numbered 4 from the first set and the object numbered 7 from the second set. The Graeco-Latin square of order 10 shown in Figure 12 is due to

E. T. Parker, one of the men who first proved its possibility.

One notices that the nine squares forming the lower, right-hand corner of this array constitute a miniature Graeco-Latin square of order 3 in the two sets of symbols $(7, 8, 9)$. In fact it is the same as the one shown in Figure 11. All squares of order 10 so far constructed have the property of containing within themselves a subsquare of order 3. Why this happens, or even whether it always happens, is not known.

□

Here is an apparently innocent programming problem that can lead to hours of investigation. What is the longest finite sequence of 1's that can be "printed out" by three cards, each with three sets of instructions? The instructions tell what is to be done under three possible contingencies and give directions how to proceed to the next card. Also in order to make the sequence finite, one stop instruction must appear somewhere.

Refer to Figure 13, and let us look at card No. 1. The top line says if we find a blank space

[13]

	Card 1			Card 2			Card 3
B	1–R–2		B	1–L–2		B	1–L–3
0	1–L–3		0	1–L–1		0	1–L–3
1	B–R–2		1	O–R–1		1	1–R–STOP

(B), print a 1, move to the right, and go to card 2 for instructions. The second line says if we find a zero (0), erase it, print a 1 in its place, move to the left, and go to card 3 for instructions. The third line says if we find a 1, print a blank (that is, erase the 1 and put nothing in its place), move to the right, and go to card 2 for instructions.

Suppose now we start our "machine" by feeding in the three cards of Figure 13 so that it first reads card 1. We have nothing on the output, so all spaces are blank, and hence it takes the instruction opposite B, which is to print 1, move one space to the right, and go to card 2 for instructions. The output register now looks like this, where we purposely did not start at the left-hand end in case an instruction asked us to go to the left:

		1							

When we have moved to the right we again encounter a blank, for which card 2 says to print a 1, move to the left, and go to card 2 for instructions:

		1	1						

On card 2 opposite 1 it says erase, print a zero, move to the right, and go to card 1 for instructions. Instead of erasing, we wish to keep a record of all operations, so we put a line through the 1 and write the new 0 beneath:

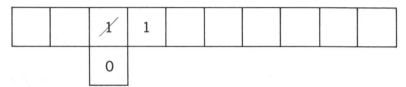

Card 1 says when you find a 1, erase, move to the right, and go to card 2 for instructions:

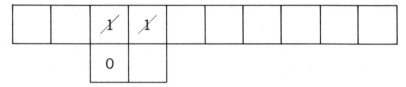

You should now be able to continue the routine yourself. The sequence will finally lead you to the last line of card 3, after which the completed print-out will be the contents of the bottom squares of all columns:

The original problem was to design a program that would print out the sequence of 1's of maximum possible length. The late Tibor Rado, who invented the game, had been unable to discover any 3-card program of this type that would produce a string of more than seven 1's, and his 7-program is the one reproduced here. In 1968 David S. Hough, while an undergraduate at Swarthmore College, devised two 8-programs and a 10-program. This last, given in the Notes, as far as I know still holds the record. Very little is known of any theory connected with the problem. It is entertaining yet perplexing and probably fundamental mathematics.

☐

A weighing problem that has been going the rounds of some mathematical communities is the problem of the balls. Given n distinguishable balls, of which it is known that no two have the same weight, it is required that they be arranged in order of magnitude by weighings of one against one in a pan balance scale (no weights). What is the smallest number of weighings, as a function of n, that will always suffice? What strategy (procedure) does one adopt to achieve this minimum? Some assortments (depending on luck) will be more quickly ordered than others. Will the best strategy automatically minimize the *expected* number of weighings, if the original ordering is random?

□

"The Theory of Games is a method of analyzing a conflict, according to the following abstraction: The conflict is a situation in which there are two sets of opposing interests; it may be regarded as a game between two players, each of whom represents one set of interests. Each player has a finite set of strategies from which he may, on any given play of the game, choose one. The total assets of the players are the same at the end of any one play of the game as at the beginning." Of course some of them have changed hands, from one player to the other; but no assets have leaked away, evaporated, or been paid to the house. "Each player wishes to pursue a conservative plan which will maximize his average gains; these maximum average gains, called the value of the game, may be calculated. Each player can, through proper play, be sure that he will receive the value of the game; to ensure this, he must choose his strategy properly—and a method exists for deciding which strategy to choose."

A strategy means a fixed procedure which one player adopts against the opponent or nature or chance or whatever device the "other player" can be considered to be. John von

Neumann, who virtually invented game theory, has discussed thoroughly the theory of two-person, finite, zero-sum games. Sometimes there is no single best strategy; but von Neumann's main theorem is that in all simple two-person games there always exists a certain *mixed* strategy which is optimal. That is, one sometimes plays one way and sometimes another, in certain definite ascertainable proportions.

Game theory is a new and little-explored field, with many applications. Much work remains to be done. Although two-person games have been investigated, less is known of the theory of n-person games where $n = 3$ or more.

4

Geometrical problems

There exists a large class of pure geometry problems of which the following is a typical example.

"In a tetrahedron $ABCD$ let L be the second Lemoine point (i.e. the point whose distances from the planes of the faces are proportional to the circumradii of these faces) and let L', L'', L''' be the harmonic conjugates of L with respect to the points where lines through L cut the edges BC and DA, CA and DB, AB and DC. Show (1) the tetrahedron $LL'L''L'''$ is self-conjugate with respect to the sphere $ABCD$; (2) the polar planes of L, L', L'', L''' with respect to the sphere $ABCD$ coincide with the polar planes of these points with respect to the tetrahedron $ABCD$."

It would take considerable effort and a

lengthy discussion simply to lay the ground-work necessary to understand the problem. There is a small group of enthusiasts who still delight in this field of pure synthetic geometry; but it is an indoor sport which was more popular in the nineteenth century than it is today.

A somewhat different kind of problem is the following. "Consider a parabola having its vertex at a variable point M on a given plane curve, and its focus at F, the point dividing the radius of curvature MC in a constant ratio; the parabola touches its complete envelope at M and also at two other finite points. The corresponding chord of contact is perpendicular to the line joining M to the midpoint of the radius of curvature at C of the evolute of the given curve. If this chord of contact intersects MF at D, then $DC = MF$." The problem is to prove these statements. At last report both this and the preceding problem remained unsolved.

This one is probably best attacked by analytic methods, removing the "pure geometry" objection if it is an objection. But you have doubtless already sensed another weakness. The problem describes an intricate set of conditions which, if fulfilled, produce an extremely specialized result that does not seem worth the top-heavy structure of the requirements. One doubts whether it could be extended or generalized. It is perhaps unfair to judge a problem without first solving it and thus discovering its inner beauty; history has many times shown the rashness of prejudging the worth of any mathematical offering. Without further comment we shall leave these two problems and move on to some others that have either the advantage of greater simplicity or the attraction of wider application.

□

Modern geometers concern themselves with many topics never mentioned in high school, representing an entirely different approach (and category of difficulty) from that of classical Euclidean geometry. You might have trouble recognizing some of these topics as geometry at all. The concept of *least covering area* yields some very ticklish questions. Here is a long-standing problem credited to Lebesgue: What is the size and shape of the least area that will cover (in the sense of a sheet of paper) any arbitrary area or point-set in the plane the maximum distance between any two points of which is 1? The covered figure can measure not more than 1 in any direction; the covering figure must be able to cover *any* such.

The circle of diameter 1 will not do the trick. Figure 14 shows an area bounded by three circular arcs of radius 1, one of the *curves of constant width* which are not themselves circles. No two of its points are separated by

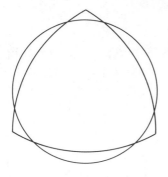

a distance greater than 1; but the circle fails to cover it. The closest approximation to date seems to be the area of Figure 15, due to R. Sprague. Starting with a regular hexagon circumscribed around a circle of diameter 1, cut from it two isosceles triangles *ABC* and *DEF* whose bases *AC* and *DF* are tangent to the circle. With *D* as center draw arc *GH*, tangent to the side of the hexagon at *G*, and a similar arc *HK* centered at *C*. The area of the corner outside the two arcs *GH* and *HK* can now be cut away.

The Lebesgue problem in three dimensions

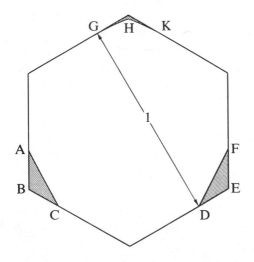

also remains open. Meschkowski gives a procedure for obtaining a body of volume .768 that covers (surrounds) all bodies of diameter 1; but this is not best possible.

Similar to the two-dimensional Lebesgue problem is the problem of the mother worm's blanket. The baby worm is one inch long. What is the shape and area of the smallest blanket with which the mother worm can surely cover the baby worm, which is not supposed to squirm once it has settled down to sleep? In less picturesque but more accurate language, what is the minimal cover for a plane curve of unit arc-length? It is known that a semicircle of diameter 1 will do, but that this is not best possible.

□

A body is *convex* if it contains all line segments determined by pairs of its points. Every plane convex region R can be inscribed in (covered by) some quadrilateral Q such that, if A_R is the area of the given region and A_Q is the area of the quadrilateral, then $A_Q \leqslant \sqrt{2} \, A_R$. Q is of course not universal; each R (even of unit diameter) may require a different Q. It is not known whether the constant $\sqrt{2}$ is best possible or can be reduced. That is, does there exist a region R for which the area of the smallest possible surrounding quadrilateral actually equals $\sqrt{2} \, R$?

The maximum distance between any two points of a plane figure is called its *diameter*. What is the plane figure of least area with a given circumference and diameter? This is a question from the theory of convex bodies, a subject in which even the elementary theory turns out to be unexpectedly intricate. A close alliance with the analytical theory of inequalities invests the topic with added interest. We shall describe a few more of its unsolved problems.

Given a plane convex figure F with two perpendicular chords that cut its perimeter into four equal parts, then twice the sum of the lengths of the chords is thought to be at least equal to the perimeter of F. Equality holds only for rectangles. Another guess about this same figure is that the sum of the lengths of the chords is at least the diameter of F. Neither of these conjectures has been proven.

A circle is a plane curve such that all chords through a given fixed point (in this case the center) have the same length. This property, however, does not characterize a circle: there are other convex curves with the same property. It is not known whether there exists a plane curve, convex or otherwise, such that two points, A and B, both have the equichordal property; that is, all chords through A or B are equal.

If the perimeter of a simple closed curve is bisected by every line through some fixed

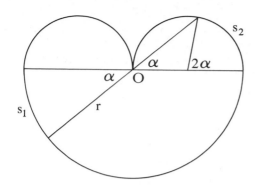

[16]

point O, is the curve necessarily centrally symmetric with respect to O? The answer is no, as shown by Figure 16, which consists of three semicircular arcs. Circumferences vary as radii, so that the horizontal line bisects the total perimeter; and for any other line through O, $s_1 = \alpha r$, $s_2 = 2\alpha \cdot r/2 = \alpha r$, where α is measured in radians.

Murray Klamkin asks the interesting question, Is the answer yes for *convex* closed curves?

A homogeneous solid cylindrical rod whose cross section is a circle has floating equilibrium (if it is light enough to float at all) in any position; that is, no matter how we turn it around its axis, it will remain at rest on the surface of the water in that position. There are curves other than circles with this property; but in three dimensions the problem is still unsolved. If a convex body floats in equilibrium in any orientation, it is necessarily a sphere?

There are associated problems. Hallard T. Croft suggests this procedure: Denote plane sections of a convex body as follows:

V: those that cut off a certain constant volume;

P: those whose section has a certain constant area;

S: those that cut off a certain constant surface area of the body;

T: those making a constant angle with the tangent planes to the body at all points of the boundary.

Then ask questions by pairing letters; for instance, one could ask, if all sections of type V are also of type S, is the body necessarily a sphere? I do not know which, if any, have been answered.

If B is a convex body, consider all the sections of B parallel to a fixed plane P. Let the maximum of these areas be called the HA-measurement of B relative to P. Now choose a new direction for plane P and find the corresponding HA-measurement of B. If all the HA-measurements are the same, regardless of the position of P, is B necessarily a sphere? The answer to this question is unknown, although the answer for non-convex bodies is no. Another question is, if the HA-measurements are not constant, would full knowledge of all of them determine B? In other words, is a convex body characterized by its HA-measurements?

□

Geometrical problems

Some investigations have been made of the theory of inscribed plane figures that can be rotated freely in a circumscribing polygon while remaining in contact with every side of the polygon. Michael Goldberg, of Washington, D.C., an expert on this topic, asks, "Are there non-spherical shapes which can be rotated through all orientations while remaining in contact with the three faces of a regular triangular prism?" Such questions, which seem very abstract, find application to problems in mechanics.

☐

Imagine a two-dimensional room whose walls form a simple closed figure, and that these walls are completely covered with mirrors. If the room were a convex body every point in it would be illuminable by a single lamp or candle placed anywhere in the room, with or without the mirrors. But suppose it is not convex. Is there a shape for the curved walls such that from some point in the room a light source would fail to illuminate the whole room even with the aid of the mirrors? The answer is yes; and more surprisingly perhaps, we can exhibit a room (Figure 17) not fully illuminated from *any* point in it. The area marked *green* is bounded by a semi-ellipse and its major axis through foci p, p'. The *blue* region is another

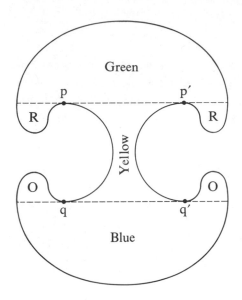

Green

p   p′

R   R

Yellow

O   O

q   q′

Blue

[17]

such semi-ellipse area, foci at q, q'. The other curves are tangent to the major axes at the foci. Now it is a property of the ellipse that any light ray crossing the major axis between the foci is reflected along another path that crosses between the foci; likewise a ray crossing the major axis beyond one focus is reflected beyond the other focus. This is because a ray emanating from a focus is reflected to the other focus, the reason for the name. Now if R stands for red points and O for orange points, we see that any light ray starting from a red point can never reach a yellow, blue, or orange point, and one from an orange point can never reach a yellow, green, or red point. A ray starting from a green point misses all the orange points, from a blue point misses all the red points, and from a yellow point misses the orange and the red points. Thus the whole region, bounded by smooth curves, is not illuminable from any of its points.

Whether such a polygonal region exists is not known, even if the polygon approximates very closely the region of Figure 17. In fact, no polygonal region is yet known that is non-illuminable from even one of its points, let alone all of them.

□

Is there a plane point-set such that every set congruent to it contains exactly one lattice point? The lattice points in the (x, y) plane are all those points both of whose coordinates are integers. The problem asks whether there exists a geometric configuration such that if this configuration were superimposed on the coordinate grid in *any position*, one and only one lattice point would be a point of the configuration. The answer is easily seen to be no for simply connected plane figures, whether convex or not. For let d be the diameter of such a figure; then if $d < 1$ the figure can, in some position, surely fail to cover any lattice point, whereas if $d \geqslant 1$ it can be placed to cover at least two lattice points. One must seek a more bizarre point-set. If there is a solution it might consist of a great many, possibly an infinite number, of disjoint pieces, or perhaps be an everywhere discontinuous set.

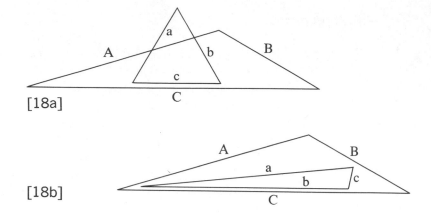

[18a]

[18b]

Steinhaus asks, when can a triangle be fitted inside another triangle? A given triangle has sides of lengths A, B, and C, and another has sides of lengths a, b, and c. What conditions connecting the six numbers A,B,C,a,b,c are necessary and sufficient to guarantee that the second triangle can be placed inside the first? At first glance it might be thought that the answer is, if and only if $a < A$, $b < B$, $c < C$. But these inequalities are neither necessary nor sufficient. In Figure 18a, $a < A$, $b < B$, $c < C$ and yet abc is not covered by ABC. In Figure 18b, $a > A$, $b > B$ and yet abc is covered by ABC.

Here is a difficult problem of solid geometry. Given a tetrahedron $ABCD$, let P be the point such that $PA + PB + PC + PD$ is a minimum. Find the conditions for P to lie inside (not on the boundary of) the tetrahedron.

A circle C has the property that every point of C is the vertex of a square all of whose vertices lie on C. Other plane curves are known to have this property. Question: Is the circle the only such curve of constant width? Another question: Does every simple closed plane curve contain at least one set of four points that are the vertices of a square? "Contain" means *on* the curve, not within the area bounded by it; nor is it required that the whole square lie within the region bounded by the curve. The non-convex curve of Figure 19 complies with the conditions. I suspect that there are some sufficiently nasty closed curves no four of whose points are the vertices of a square, but no one has succeeded in producing one, nor even in proving its existence.

It is known that there can be no equilateral triangle all of whose vertices are lattice points in the plane. In the space of ordinary three-dimensional analytic geometry, all the points each of whose coordinates are integers form what is called a "cubic lattice." It is known that there do exist equilateral triangles all of whose vertices are points of a cubic lattice. The problem is to characterize them; that is, determine (specify) the entire class of such triangles.

□

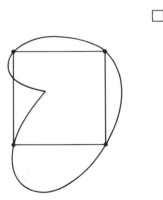

If the edges of a convex polyhedron are all tangent to a sphere of unit radius so as to form a "crate" from which the sphere cannot escape, what is the minimum possible total length of the edges?

What is the regular m-gon of greatest area that can be inscribed in a regular n-gon? "Inscribed" in this instance means only that the inner figure must be contained wholly within or on the boundary of the outer figure. What is wanted is a formula (or formulas) for various values of m and n, especially for $m > n > 4$.

For a convex curve, there always exist three concurrent chords bisecting each other and intersecting at angles of $60°$. Steinhaus, who proved this, believes it to be true for any simple closed curve, whether convex or not, but he has been unable to prove it.

It is impossible to bend a closed convex surface, a fact exploited by nature in the design of eggs, whose shells would have to be far heavier if the egg had any other shape. When a ping-pong ball is dented, some tearing or stretching of the surface actually takes place. If an arbitrarily small hole is cut out of a convex surface, the surface can then be bent. It is not known whether it is sufficient merely to slit the surface, or whether even the removal of some isolated points would make it bendable.

□

Geometrical problems

A very slippery question in plane analytic geometry is, if the six conics determined by each five of a set of six points are congruent, must they coincide? Stated in 1943, the problem is still unsolved despite having received considerable attention.

How many points are there on a plane curve of second degree (a conic) such that the distance between any two of them is rational? This seemingly geometric question actually belongs to analysis or number theory. There has apparently been little headway made toward its solution since its proposal in 1970.

Two lines in the plane in general intersect in one point. The exceptional case is parallelism, when they do not intersect at all. In three dimensions the situation is somewhat reversed: two lines do not intersect (they are said to be skew to each other) unless they lie in the same plane, in which case they do intersect unless they are parallel. Two skew lines have one and only one common perpendicular, the length of which measures the shortest distance, or more simply the distance, between the two lines. How many lines can be drawn in 3-space, each a unit distant from every one of the others? It is conjectured that seven is the maximum number, but no proof is available. Seven might be too high or too low.

□

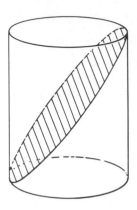

[20]

What is the largest plane section of a solid right-circular cylinder of radius r and altitude h? It is not, as might first be supposed, the elliptical section obtained by slicing diagonally (Figure 20). A section nearer the vertical, cutting across the ends of the cylinder, has greater area. The angle of the cutting plane for maximum area depends on the ratio of h to r. One can set up an expression for the area as a function of this ratio and apply the maximizing device of elementary calculus; but the resulting equation is awkward and can probably be solved only by methods of approximation. A formula for the maximum area in terms of h and r is not at present accessible.

□

Problems of dissection owe their difficulty to the fact that they follow no apparent pattern. Every solution is different, and each problem must be handled separately.

Here is one that seems relatively easy, and might even have a solution before this book appears. When can a non-right triangle be cut up into 5 triangles, not necessarily congruent to each other, but all *similar* to the original triangle? It is known that any triangle is dissectable into n triangles similar to itself when $n = 4$ or $n \geqslant 6$. If $n = 2$ or 3, only a right triangle is so dissectable. If $n = 5$, a right triangle is so dissectable; but what about a non-right triangle? This was the only case left open in 1971.

A square can be cut into 24 smaller squares all of different sizes. (See Figure 21.) Is this the

[21]

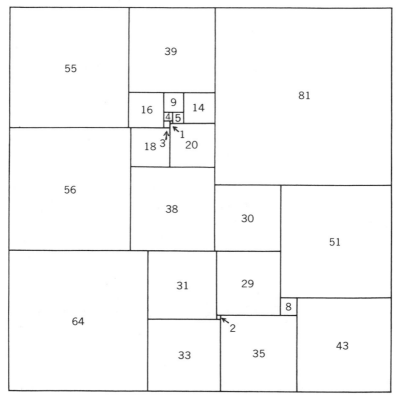

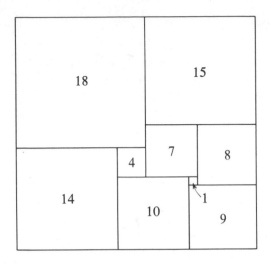

[22]

least possible number, if no two squares are to be alike?

Figure 22 shows a dissection of a 33 × 32 rectangle into 9 non-congruent squares. A dissection into squares, no two of which are equal, is called *simple* if it does not contain any sub-rectangle or square. Thus although a dissection of a rectangle of proportions 2:1 could be achieved by conjoining two different 25-element square-dissected squares (of which ten are known), it would not be simple. In 1970 a simple dissection into 23 squares was found for a 2 × 1 rectangle. Its total side length is 270 × 135. Best possible dissections for most rectangles are unknown, nor do we know which rectangles have simple dissections into a reasonably small number of squares.

The question has been asked, is the sum of the first k perfect squares ever itself a square?

The answer has long been known to be, "Yes, but only once: $1^2 + 2^2 + 3^2 + 4^2 + \cdots + 24^2 = 4900 = 70^2$." It is curious that exactly 24 squares are required for the dissection of Figure 21. One wonders, naturally, whether the *first* 24 squares could possibly be assembled to form a 70 × 70 square. It is almost certain that this cannot be done, but no proof has been produced to date. In fact, no square smaller than that of Figure 21, which has side 175, has been discovered.

If we remove the restriction that the sub-dividing squares must all be of different sizes, the problem seems easier. Consider $n \times n$ squares, n an integer. We agree not to allow the "trivial" subdivision of the 4 × 4 square that merely mimics that of the 2 × 2. (Figure 23.) In order to impose this agreement, we require that there be no common factor of the side-lengths of all of the subdividing squares. Such a subdivision is called a prime dissection of the square into squares. Figure 24 shows minimal prime dissections for $n = 2$, 3, and 4. If $S(n)$ means the smallest number of squares that can subdivide an $n \times n$ square with the prime re-

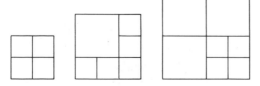

striction, then we have $S(2) = 4$, $S(3) = 6$, $S(4) = 7$. What we would like to have is a procedure for doing the subdivision in general, and a formula for $S(n)$ in terms of n. This has been called the problem of Mrs. Perkins' quilt, and as Martin Gardner succinctly puts it, the solution is nowhere in sight.

☐

The present state of knowledge concerning various problems of dissection and reassembly is displayed in Figure 25. The numbers in the

[25]

Regular polygon of N sides; N =								
4	4							
5	6	6						
6	5	5	7					
7	10	9	11	11				
8	8	5	9	9	13			
12	8	6	.	.	.	.		
Greek cross ✚	5	4	7	7	12	9	6	
Latin cross ✝	5	5	8	6	12	8	7	7
	3	4	5	6	7	8	12	Greek cross ✚
	Regular polygon, No. indicates No. of sides							

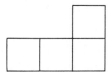

boxes indicate the least known number of pieces required to dissect the plane figure listed on the left and reassemble (turning pieces over if necessary) to form the indicated figure at the bottom. This table gives the current (1971) minimum figures. In nearly all cases it is not known whether these are best possible.

Dissection problems in three dimensions can be difficult. Given a tetrahedron, can it be divided by plane cuts into a finite number of pieces that can be reassembled to form a cube? It is known that some tetrahedra can be so dissected, but the list is not complete.

□

The theory of polyominoes has appeared within the last decade. A polyomino is to be imagined (or actually cut out of a piece of cardboard, wood, or what have you) to be made up of a certain number of squares joined along one or more edges. Figure 26 shows a tetromino, and Figure 27 two pentominoes. Of course there are many others.

The game of making patterns with these

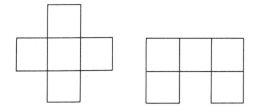

objects contains many difficult problems. For example, there exist twelve differently shaped pentominoes. A man wishes to construct a set of one of each of them from a single piece of plywood, but his saw will not cut around corners. What is the smallest plywood rectangle he can buy? It is known that it can be done from a 6 × 13 piece; but this wastes eighteen squares, because twelve pentominoes actually occupy only sixty squares. Is there any more economical way of making the cuts?

It is known that there are 12 pentominoes 35 different hexominoes, and 108 different septominoes. But no one has succeeded in finding a general formula for the number of n-ominoes in terms of n.

Tiles in the shape of regular hexagons are sometimes used for tiling bathroom floors (see Figure 44). There are also many non-regular polygons with the necessary property: identical (congruent) shapes endlessly repeated can cover the plane. Any triangle will do it. So, for that matter, will any quadrilateral.

A further requirement produces a new set of unsolved tiling problems. Suppose that not only must the tiling be accomplished by congruent figures, but also that these figures must each be decomposable into exactly similar replicas congruent to each other. Figure 28

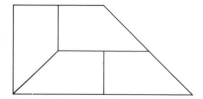

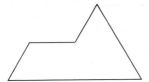

shows such a shape. Each smaller quadrilateral could again be decomposed, and so on indefinitely. It follows that by reversing the process any such figure must be capable of tiling the plane. These replicating figures have acquired the name "rep-tiles." (Sorry about that; I didn't invent the name.)

The only known rep-4 pentagon, called the Sphinx, is shown in Figure 29, and it is a nice problem to locate the decomposition into four similar pieces. The following are unproven conjectures. (1) Every rep-tile also tiles a parallelogram. (2) The only rep-2 figures are the isosceles right triangle and the parallelogram whose sides are in the ratio of $\sqrt{2}:1$ (Figure 30). (3) A rep-tile with five or more sides cannot be convex; the Sphinx, for example, is not.

One can also do certain tilings with polyominoes. Can a rectangle be tiled with the hexomino of Figure 31a? With the heptomino of Figure 31b?

In 1966 a different kind of tiling problem was suggested. It is known that the plane can be tiled with an infinite number of triangles no two of which are similar, all with all three sides

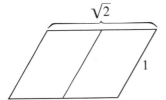

$\sqrt{2}$

1

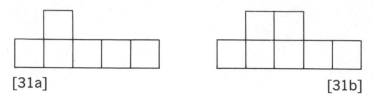

[31a]   [31b]

rational. What is not known is whether such a tiling can be accomplished with triangles all of whose sides are integers.

□

A long-standing problem deals with the closest *packing* of spheres of equal sizes. Usually one imagines the spheres packed in a small part of a very large container, so that border conditions are neglected. The closest packing known has a density of 0.74; that is, nearly three-quarters of the space is filled by the spheres. Whether this is the densest possible is not known; but it was proved in 1958 that if there exists a closer packing its density cannot exceed 0.78. This is a theoretical absolute upper limit.

A much simpler question is the problem of the Twelve Pennies. Twelve equal disks (pennies) can be packed, non-overlapping, in a 3×4 lattice-point array (Figure 32). The disks then cover $\pi/4$ of the area of the rectangle, for if each disk has diameter 1, its area is $\pi r^2 = \pi/4$, and it lies in a square of unit area.

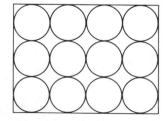

[32]

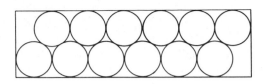

The unproven conjecture is that this is best possible; that is, that the 12 disks can be packed into no smaller rectangle. The 2×6 lattice arrangement is no better and no worse. A candidate for "next best" seems to be the 2×6 skew-lattice arrangement of Figure 33; but its area is slightly greater than 12.

Concerning the packing of squares we mention two tantalizing little problems. (1) What is the smallest number S such that any set of squares of total area 1 may be packed in a rectangle of base 1 and height S? It is at least $\sqrt{3}$, to accommodate 3 squares each of area $\frac{1}{3}$; but it is not known whether $\sqrt{3}$ is the smallest possible. (2) What is the smallest number T such that any set of squares of total area 1 may be packed in some rectangle of area T?

The following problem has been (partially) solved in its plane version. Let A and B be equal disks. If A is cut by a chord into two pieces, A_1 and A_2, what is the smallest square in which A_1, A_2, and B can be packed? Because the chord is arbitrary, the solution varies with the position of the cut. Ideally one would like to have a formula connecting the length of the side of the square with the length of the chord. In three dimensions the problem has not, to my knowledge, been attacked: Given two equal spherical balls, one of which has been sliced into two pieces by a plane, what is the smallest cube that will contain these three convex bodies?

☐

An unspillable saltcellar always returns to an upright position, by virtue of its weighted base. Are there polyhedra that will stand on only one face, even though the polyhedron is to be considered as a solid of uniform density? The answer is yes. Figure 34 shows a unistable polyhedron of 19 faces, standing in its only stable position. It is a 17-sided symmetrically truncated prism, the planes of the truncating faces meeting at an angle just wide enough so that the solid will not stand on either end. The vertical cross section through the center is designed as indicated in Figure 35, the right triangles being similar. Richard K. Guy, who devised this ingenious model and proved that

[34]

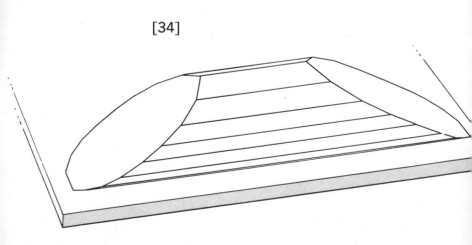

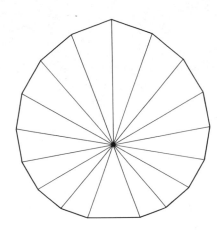

it is unstable, proved also that no fewer than 17 lateral faces would do. In the same paper he also poses several new problems. In all cases density is assumed to be uniform.

(1) Does there exist a unstable polyhedron, of some pattern different from this one, with a total of fewer than 19 faces?

(2) It is known that any tetrahedron is stable on at least two faces. A first step toward closing the gap between 4 and 19 would be to show that every pentahedron is stable on at least two faces. Even this is not known.

(3) Guy's 19-hedron has two axes of symmetry. Can a unstable polyhedron have more than two?

(4) Can the resting face of a unstable convex polyhedron be the face of least diameter? (In the 19-hedron it happens to be the face of greatest diameter.)

□

We usually think of 2-space (a plane) as the space generated by two directions at right angles to each other, or two orthogonal axes. Similarly 3-space, the space in which (we think) we live, is generated by three orthogonal axes. The mathematician does not hesitate to talk about 4-space (four orthogonal axes), or n-space. I suggest that, despite the claims of some people, it is impossible to visualize or adequately sketch a figure in 4-space. We have to collapse 3-space into a plane in order to make a perspective drawing of a solid figure. It is only because our eyes do this trick for us every moment of our waking lives that it has become easy. If you close one eye (to eliminate binocular depth perception), what you see is exactly what the camera sees—a flat picture. Only a lifetime of unconscious practice gives this picture its third dimension.

To make a drawing of an ordinary three-dimensional figure, we have to reduce the number of dimensions by only one, but in any attempt to draw a four-dimensional figure, two dimensions are lost. It is analogous to trying to describe a cube by noting the distribution of 8 points on a *single straight line*. In Figure 36, A is the cube in two dimensions and

[36]

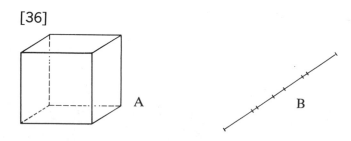

A   B

B is the same cube drawn in one dimension. I submit that B is not a very meaningful picture of a cube, nor does it give an idea of what a cube "really looks like." So what hope have we of indicating what a four-dimensional object "really looks like," if that statement has any meaning, by means of two-dimensional diagrams?

If we don't stop here we will quickly get into difficult semantic and philosophic arguments. What should be emphasized is that mathematicians have no interest in such arguments, and are perfectly content to work in Euclidean n-space, where a vast amount of mathematics takes place, without attempting to visualize anything. Very many geometric problems can be generalized in n-space.

What is the shortest path, within a cube, from one point to another on the boundaries of the cube and touching certain specified faces along the way? The problem has been extended and in fact solved in some generality in 4-space, but not in n-space for $n \geqslant 5$. Also the extension to figures other than cubes, even in 4-space, is wide open.

A *polytope* is an n-dimensional polyhedroid. As a polyhedron has one more dimension than a polygon, so polytopes have more dimensions than polyhedra. In 1967 Branko Grünbaum wrote a 450-page book entitled *Convex Polytopes*, which is packed with unsolved problems. So rapid is progress in this popular sector of mathematics that he wrote in 1970, "Many parts of the book . . . are by now completely out of date. Even the most recent survey [1969] . . . has been overtaken in many directions by new results. . . . Hence it seems worthwhile to prepare an up to date summary of the new results." Then follows a 50-page article, accompanied by references to more than 300 other papers, most of them published very recently. The survey was submitted for publication in May 1970. Yet it was necessary for the author to append this note: "(Sept. 15, 1970.) During the few months since the manuscript of the present paper was completed, quite a few rather remarkable achievements were published, or brought to my attention." Sixteen more references are added. Presumably by the time this book appears, the late 1970 summary will again be out of date.

□

If equilateral triangles are "inscribed in" an equilateral hyperbola, the locus of centroids of these triangles is another equilateral hyperbola. Are there any other curves with this property? What is the equation of the most general curve such that the locus of centroids of inscribed equilateral triangles is the same curve?

The perpendicular bisectors of the sides of a given quadrilateral Q_1 form a quadrilateral

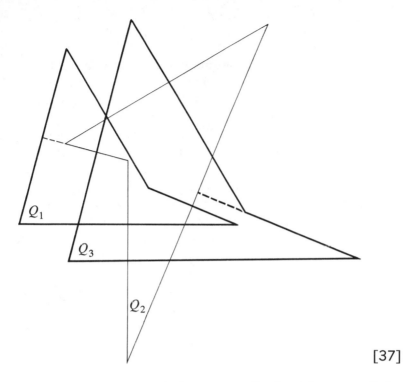

Q_1

Q_3

Q_2

[37]

Q_2, and the perpendicular bisectors of the sides of Q_2 form a quadrilateral Q_3. Show that Q_3 is similar to Q_1, and find the ratio of similitude. It is of course *not* sufficient to show that the sides are respectively parallel, which is trivially evident. An interesting feature is that the construction is not reversible: if you start with Q_3 (see Figure 37), you do not return to Q_1. It is easy to see why not. The sought ratio of similitude is greater than one, in the quadrilateral which we happen to have chosen; hence the new quadrilateral must always be larger than the old. In fact if the construction is iterated, continuing the same notation we observe that Q_4 and Q_2 bear the same relation as Q_3 and Q_1, and this property is general.

Our catalog of geometric questions has finally led us back to problems with which a classical geometrician would feel at home. We close this chapter with another problem from ordinary, old-fashioned plane geometry.

Through the vertex A of a triangle ABC a straight line AM is drawn cutting the side BC in M. Let 2θ be the angle AMC; O and I the centers of the circumscribed circle (O) and the inscribed circle (I) of ABC. The circles (ω_1) and (ω_2) with centers ω_1 and ω_2 and radii ρ_1 and ρ_2 are each tangent to (O) and the first is tangent also to the two sides of angle AMC while the second is tangent to the two sides of angle AMB. Prove that: (1) The straight line joining ω_1 and ω_2 passes through I. (2) The point I divides the segment $\omega_1\omega_2$ in the ratio $\tan^2\theta/1$; and $\rho_1 + \rho_2 = (r \sec \theta)^2$, where r is the radius of (I).

The problem is elementary but difficult. Part 1 can almost surely be solved by someone with sufficient ingenuity and persistence using only synthetic (Euclidean) methods. Part 2 is probably more amenable to methods of analysis and co-ordinate geometry.

This problem was proposed by Victor Thébault of Le Mans, France, and published in the American Mathematical Monthly in 1938. It is one of hundreds submitted by this most productive problemist between 1933 and 1960 when he died at the age of seventy-eight. He was adept at number theory as well as geometry. "M. Thébault's prolific output of theorems and problems about numbers is a source of constant admiration, little short of wonder, on the part of all those upon whom the higher arithmetic exerts its fascination." M. Thébault's own views are worth quoting: "Some mathematicians exhibit a tendency, not altogether free from a certain disdain, to see in such problems only insignificant trifles. Trifles, if you please, but the solution of which often demands no less penetration of mind, ingenuity, and subtle artifice than many questions of allegedly deeper significance. Moreover the study of an elementary proposition sometimes necessitates an effort which is far from negligible, which constitutes an excellent intellectual exercise, and which leads to something worthwhile indeed."

5

Arithmetical problems

Number theory, sometimes called the higher arithmetic, abounds in unsolved problems of a very considerable degree of difficulty. The field represents an example of what we discussed in Chapter 1: although extremely difficult by numerical methods, some of the problems have recently been solved by embedding them in the field of analysis. Since number theory deals exclusively with the properties of *whole numbers,* the integers 1, 2, 3, 4, . . . , it was an unexpected discovery that any results at all could be obtained through methods usually reserved for *continuous* operations. In analysis one deals with variables that can take on any real value whatever, including the integers but also including all the (infinite supply of) numbers between the integers. Most of the recent results have been obtained by analytical methods of an advanced nature and it is a reasonable hope that further progress will be made in that direction.

Perhaps the most famous unsolved problem in all of mathematics is Fermat's Last Theorem. It should be called a conjecture inasmuch as Fermat did not leave us a proof. He said he had a "wonderful" proof; and in every other case where he made such a statement a proof was subsequently found, but never for the last theorem.

The equation $x^2 + y^2 = z^2$ has many solutions, in fact an infinite number of them, in which x, y, and z are all positive integers. A familiar one is $3^2 + 4^2 = 5^2$, and another is $5^2 + 12^2 = 13^2$. It is easy to write formulas producing all the solutions. In fact these formulas, or their equivalents, are rediscovered every year by vigorous amateurs. To discover something oneself is no less exciting or praiseworthy because it has been done previously by someone else.

Now consider the equation $x^n + y^n = z^n$. Fermat's Last Theorem states that this equation has no solution whatever in positive integers if n is any integer greater than 2. This most surprising fact—and it does seem to be a fact, even though no general proof has yet been produced—has been checked for all n up to 5500. The checking can be done by refining the problem and feeding it into a large electronic computer. Inasmuch as even a small number like 2 raised to the five-thousandth power yields a perfectly enormous integer, there is no possible chance that anyone can stumble upon a counterexample to the theorem by hand methods: there are none within reach. What remains to be done with pencil and paper and brain power is precisely that which cannot be done by any machine: to find a proof covering at once *all* cases, so that it is no longer a matter of trial and error. Fermat said he possessed such a proof. The unsuccessful search for it by hundreds of competent mathematicians over the years has led many to believe that Fermat was mistaken, that what he thought was a proof—and no one questions his sincerity—must have contained a flaw. Whether that is so or not, the search has been far from useless. From it have stemmed many other interesting results. A branch of mathematics called the theory of algebraic numbers was developed largely as a consequence of this very search.

Consider the patterns exhibited below.

$$n(n + 1)$$
$$\downarrow$$
$$1 + 2 = 3$$
$$4 + 5 + 6 = 7 + 8$$
$$9 + 10 + 11 + 12 = 13 + 14 + 15$$
$$16 + 17 + 18 + 19 + 20 = 21 + 22 + 23 + 24$$
$$\cdots \cdots \cdots \cdots = \cdots \cdots \cdots$$

$$[2n(n + 1)]^2$$
$$\downarrow$$
$$3^2 + 4^2 = 5^2$$
$$10^2 + 11^2 + 12^2 = 13^2 + 14^2$$
$$21^2 + 22^2 + 23^2 + 24^2 = 25^2 + 26^2 + 27^2$$
$$36^2 + 37^2 + 38^2 + 39^2 + 40^2 = 41^2 + 42^2 + 43^2 + 44^2$$
$$\cdots \cdots \cdots \cdots = \cdots \cdots \cdots$$

Both these arrays can be continued indefinitely. Their laws of formation are strikingly similar. The column of figures immediately to the left of the equal signs is the key to both. In the first array the numbers 2, 6, 12, 20, . . . are of the form $n(n + 1)$, with $n = 1, 2, 3, 4, . . .$ In each equation there are $(n + 1)$ consecutive integers on the left and n on the right. In the second array the numbers 4, 12, 24, 40, . . . are just twice $n(n + 1)$, and this time it is the consecutive *squares* adjacent to $[2n(n + 1)]^2$ that are displayed according to the same prescription.

Seeing so much in common to the two arrays we should expect the pattern to be extendable still farther—but it is not. The next array ought to center around values of $[3n(n + 1)]^3$, but we know that it cannot. For if the system were extendable to cubes our first equation, corresponding to $n = 1$, would be

$$5^3 + 6^3 = 7^3,$$

which Fermat's Last Theorem tells us is impossible.

The interesting point is that this constitutes a very near miss.

$$5^3 + 6^3 = 341$$

and

$$7^3 = 343.$$

If someone had deliberately set out to tease us, he could hardly have done a better job.

What is it that allows the equation

$$x^n + y^n = z^n$$

to have an infinite number of integral solutions for $n = 1$ and $n = 2$, and none for higher n? This is the question that has baffled mathematicians for centuries.

□

Number theory frequently concerns itself with primes; in fact one could almost say that number theory *is* the study of prime numbers. A prime, you will recall, is a number like 5 (or 11 or 29) which has no divisors except itself and 1. A number like 15 is said to be composite: its factors are 3 and 5. Primes are special numbers, for many reasons, and we should therefore like to know all about them. At present we know almost nothing about them. It is known that they constitute an infinite set: there is no last prime. But what are they? How can you tell by looking at a given number, say 39,617, whether it is prime or composite? How many primes are there smaller than 39,617? What is the first prime after 39,617? What is the sequence of gaps between primes? None of these questions has ever been answered.

The only known (workable) test for the primality of a number N is to divide it by all the lesser primes 2, 3, 5, 7, 11, 13, . . . , up to the square root of N, one after the other, thus testing by sheer trial whether N has a divisor. There is no need to go beyond $\sqrt{N}$; for if a larger number divided N, the quotient would be *less* than $\sqrt{N}$, and hence would have been found in a previous trial.

There is no known formula that generates all the primes; indeed, there is no formula that guarantees to produce *any* of them. Fermat thought he might have such a formula, but it is important to note that this time he confessed that he could find no proof, which is hardly surprising inasmuch as the supposed formula breaks down. Fermat found that

$$X = 2^{2^n} + 1$$

produces a value of X which is a prime for the first few n's. For $n = 0, 1, 2, 3,$ and 4 respectively we obtain X's, called Fermat numbers, 3, 5, 17, 257, and 65,537, which are prime. But the next Fermat number, $F_5 = 4,294,967,297$, is factorable into $641 \times 6,700,417$. For $n > 5$ the Fermat numbers rapidly become huge. A few are known to be composite. It is not known whether any higher ones are prime.

Between 1880 and 1925, factors were obtained at great labor by various investigators for ten different Fermat numbers between F_5 and F_{73}. Twenty-eight years later John L. Selfridge knocked off two more on the SWAC, a modern electronic calculator. Then in 1956 and 1957 Selfridge and Raphael M. Robinson, on the same machine and with a fraction of the labor previously required, found factors for twenty more Fermat numbers, most of them larger by far than any that had been tackled before.

Some factors are known for a few of the very large F_n; but two of the smallest, F_7 and F_8, have interesting histories. Through the application of certain high-powered divisibility criteria it is possible to decide whether a Fermat number is composite without actually finding its factors. In 1905 it was ascertained that F_7 is composite, but there the matter stood for sixty-five years. F_7 is so large (a number of 39 digits) that it defied factorization. Finally in 1970 its two prime factors were discovered by Morrison and Brillhart, of course with the aid of a very large computer system. Once the program had been devised, the factorization required an hour and a half of running time on an IBM 360/91.

F_8 is a number having twice as many digits as F_7, and it too has been found to be composite (1909), but that is all. Its prime factors are few, large, and unknown.

□

The statement "12 is congruent to 5 modulo 7" is written $12 \equiv 5 \mod 7$ and means that 12 and 5 have the same remainder when divided by 7. Thus $31 \equiv 1 \mod 3$ and also $31 \equiv 10$

mod 3. Congruences behave like equations under many operations. One can easily prove that it is possible to multiply both sides of a congruence by the same integer, or to square both sides, without destroying the congruence. Thus: $4 \equiv 1$ mod 3. Multiplying by five: $20 \equiv 5$ mod 3; or squaring: $16 \equiv 1$ mod 3. The power of congruence arithmetic is illustrated by two examples.

(1) Every odd perfect square is congruent to 1 modulo 8. For every odd number is by definition congruent to some one of 1, 3, 5, or 7 mod 8. Squaring both sides we have that every odd square is congruent to 1, 9, 25, or 49 mod 8; but each of these is congruent to 1 mod 8, which proves the statement.

(2) What is the remainder when 3^{100} is divided by 7?

Solution:
$$3^3 = 27 \equiv -1 \text{ mod } 7$$
$$(3^3)^{33} = 3^{99} \equiv (-1)^{33} = -1 \text{ mod } 7$$
$$3(3^{99}) = 3^{100} \equiv 3(-1) = -3 \equiv 4 \text{ mod } 7$$

It may not be very valuable to know that when 3^{100} is divided by 7 the remainder is 4; but inasmuch as 3^{100} is a 48-digit number, we have obtained this information remarkably cheaply. It is only with the aid of such methods as these that one can hope to attain any results with numbers as large as Fermat numbers.

If p is a prime, 2^{p-1} is usually not congruent

to 1 mod p^2. For example,

$$
\begin{aligned}
\text{if } p = 3, && 2^2 &\equiv 4 \text{ mod } 9; \\
\text{if } p = 7, && 2^6 &\equiv 15 \text{ mod } 49; \\
\text{if } p = 11, && 2^{10} &\equiv 56 \text{ mod } 121.
\end{aligned}
$$

1093 is the first prime satisfying the congruence $2^{p-1} \equiv 1 \text{ mod } p^2$. It is not known whether there are infinitely many solutions to this congruence. Such primes are of interest in the study of Fermat's Last Theorem.

□

An easy formula that turns out a few primes is

$$X = n^2 - n + 41$$

It happens that for any n less than 41, X is a prime; but $n = 41$ yields $X = 41^2$, obviously composite. In the same way, and for the same reason, no polynomial function of n with a constant term greater than 1 can ever yield primes beyond a certain point. One might suppose that the above formula works as well as it does because 41 is itself a prime. Then one would need only to find another prime $p > 41$ by use of this formula, and set it into a new formula

$$X = n^2 - n + p$$

to produce more new primes. Inasmuch as the procedure is constructive, it would serve to generate an unceasing supply of primes—but it doesn't work. The initial supposition is false. Seven is a prime, but

$$X = n^2 - n + 7$$

is not a prime-producing formula for all $n < 7$. In fact, if $n = 2$, $X = 9$. There are many such counterexamples. There seems to be no good reason why the expression happens to produce primes up to $n = p$ when $p = 41$.

☐

The *binomial coefficients* are the numerical parts of the elements in the expanded expression for $(a + b)^n$. If we multiply out $(a + b)^2 = (a + b)(a + b)$ we get $a^2 + 2ab + b^2$. Multiplying in turn by another $(a + b)$ yields $(a + b)^3 = a^3 + 3a^2b + 3ab^2 + b^3$. Now notice the law of formation of the coefficients. If we write down the skeleton of $(a^3 + 3a^2b + 3ab^2 + b^3)(a + b)$, it looks like this:

$$
\begin{array}{ccccc}
1 & 3 & 3 & 1 & \\
1 & 1 & & & \\
\hline
1 & 3 & 3 & 1 & \\
 & 1 & 3 & 3 & 1 \\
\hline
1 & 4 & 6 & 4 & 1 \\
\end{array}
$$

From each line we can easily write the next:

$$
\begin{array}{cccccc}
1 & 4 & 6 & 4 & 1 & \\
 & 1 & 4 & 6 & 4 & 1 \\
\hline
1 & 5 & 10 & 10 & 5 & 1 \\
\end{array}
$$

This semi-intuitive argument allows us to write down as many lines as we require of the Pascal Triangle of coefficients. The nth row gives the coefficients of $(a + b)^n$ (Figure 38); to obtain any number, simply add the two diagonally above it.

In the study of prime numbers, one comes upon a curious fact, readily proved: n divides all the numbers (except the first and last) of the nth row if and only if n is a prime. Thus 5 and 7 divide each number in their respective rows, but 8 and 9 do not. We have here a theoretically perfect test of primality—but it is useless from a practical point of view, because the labor of testing the binomial coefficients of a large number would be far greater than

[38]

```
                1       1
             1      2       1
          1      3      3       1
       1      4      6      4       1
    1      5     10     10      5      1
 1      6     15     20     15      6      1
1    7    21    35    35    21    7    1
1   8   28   56   70   56   28   8   1
1   9   36   84  126  126  84   36   9   1
```

. . . .

finding out whether it was prime by some more primitive method.

This theorem disposes of prime n's; but what of composite ones? On the lines corresponding to composite n, *some* of the coefficients are evenly divisible by n and some are not. Which ones? The answer to this interesting question is not known.

Although the binomial coefficients have been the subject of intense study by numberless mathematicians, they still pose unanswered questions. We note that each number greater than 1 appears in the triangle only a finite number of times. (We except the number 1 throughout this paragraph.) There is one 2, there are two each of 3's, 4's, and 5's, three 6's, and so on. Let $N(a)$ stand for the number of times a appears in the triangle. Thus $N(6) = 3$. The following problems were unsolved in 1971. (1) Is there any formula for $N(a)$, and if so, what is it? (2) Does $N(a)$ continue to increase indefinitely with a, or is there some upper bound for N? Three is the smallest number that appears exactly twice, 6 is the smallest number that appears exactly three times, etc. (3) Is there any formula for the smallest number that appears exactly k times, and if so, what is it?

If we write the sequence of the first n primes, then the sequence of their successive differences (in absolute value), then the second differences, and so on, we obtain a difference table, like the following one for $n = 8$.

```
2    3    5    7    11    13    17    19
   1    2    2    4    2    4    2
      1    0    2    2    2    2
         1    2    0    0    0
            1    2    0    0
               1    2    0
                  1    2
                     1
```

Will the first number in all rows except the top row be 1 for all values of n? That the answer is affirmative for all n up to 63,419 has been checked by computer; but, as usual, what is missing is a proof valid for any n.

☐

How many primes are there less than a given number? No one knows, although a formula for the exact number would be most welcome. What has been known since 1896 is that the number of primes less than N approaches approximately $N/\log N$ as N becomes very large.*
This means that for large N the probability that any randomly selected integer "in the vicinity of N" should be prime approaches

---

* The symbol "log" in this book, with no base indicated, means the *natural log*, or logarithm to the base e.

1/log N, a quantity known as the *asymptotic density*.

In 1742 a man named Goldbach asked Leonard Euler whether he could prove or disprove the following conjecture: every even number greater than 2 can be written as the sum of two primes in at least one way. For example, $8 = 5 + 3$. (Sometimes it can be done in several ways: $48 = 7 + 41 = 11 + 37 = 17 + 31 = 19 + 29$.) Euler was unable to prove that this is true of all even numbers, nor was he able to find a counterexample. Goldbach's conjecture remains unsettled.

The so-called twin primes are pairs of primes whose difference is 2, like (11, 13) and (29, 31). They seem to be scattered throughout the number system. The statement that there are infinitely many of them is believed to be correct; but that is the most that we can say.

Essentially the only way to construct a table of primes up to N is by a procedure known as the "sieve of Eratosthenes." Write down all the integers up to and including N. Strike out all multiples of 2 (except 2 itself), then all multiples of 3 (except 3) which are not already gone, then all remaining multiples of 5, and so on. The reason 5 is chosen after 3 is *not* that it is the next prime (we are not supposed to know that at this stage), but that it is the next number left: 4 is already gone. In this way all the composite numbers fall through the sieve, leaving only a list of the primes $\leqslant N$.

In 1956 S. M. Ulam and others made a table of what they christened *lucky numbers* by applying a slightly different sieve. Figure 39 shows the lucky numbers less than 100. (The slant lines indicate at what stage in the construction a number was eliminated.)

"In the sequence of all integers we strike out every second one, that is to say, all the even numbers. The first number remaining (apart from 1, which will not be counted) is 3. We shall now strike out every *third* integer, counting only the remaining ones, that is to say, this time we will strike out the integers 5, 11, 17, etc. In the remaining sequence the first number not used before is 7. Therefore we shall strike out every seventh number, counting among the remaining ones again. This will eliminate 19, etc. We proceed in this manner *ad infinitum*." The numbers that remain, 1, 3, 7, 9, 13, . . . are called the lucky numbers.

"It turns out that many asymptotic properties of the prime number sequence are shared by the lucky numbers. Thus, for example, their asymptotic density is $1/\log N$. The numbers of twin primes and of twin luckies exhibit remarkable similarity up to the integer $n = 100,000$, the range which we have investigated on the machine. . . . It also happens

[39]

1	3	5̸	7	9	1̸1̸	13	15	1̸7̸	1̸9̸
21	2̸3̸	25	2̸7̸	2̸9̸	31	33	3̸5̸	37	3̸9̸
4̸1̸	43	4̸5̸	4̸7̸	49	51	5̸3̸	5̸5̸	5̸7̸	5̸9̸
6̸1̸	63	6̸5̸	67	69	7̸1̸	73	75	7̸7̸	79
8̸1̸	8̸3̸	8̸5̸	87	8̸9̸	9̸1̸	93	95	9̸7̸	99

that within the range investigated every even number is the sum of two lucky numbers."

These considerations present the primes in a new light. That so many properties hitherto thought to be sacred to the primes are shared by the luckies removes some of the charm and at the same time perhaps some of the significance of the primes. If these properties are consequences only of the fact that the primes are generated by a sieving process, and are not due to their primality, then many investigators have certainly been looking in the wrong direction. It can never be said that primes play no special role in the number system: a vast body of interesting and important theory of all kinds rests on primality. Nevertheless one somewhat reduces the preeminence of a mountain peak by discovering others of the same height.

Ulam has also considered the sequence 1, 2, 3, 4, 6, 8, 11, 13, 16, 18, 26, . . . formed according to the following rule: strike out from the sequence of natural numbers all (after 1 and 2) except those that can be obtained in one and only one way by adding two different earlier numbers of the set. Thus a number may be disqualified either because it can be obtained in too many ways, like 5; or because it cannot be obtained at all, like 23. The question is, what is the asymptotic density of this sequence?

☐

There are certain pairs of primes, such as (13, 31) and (1229, 9221), where each is the "reversal" of the other. Perhaps the palindromic primes like 151 should also be counted, being their own reversals. Are there infinitely many such pairs? If so, what is their asymptotic density?

Many *iterative* numerical problems have intriguing unsolved aspects.

(1.) Add to any number its reversal. If the sum is not a palindrome, repeat the process on the sum, and continue until a palindrome is attained. If we happen to start with the number 459, for example, only two additions are required; but 549, which does not seem very different, requires five additions. It was once conjectured that this process would always yield a palindrome for every starting number, in some finite number of steps. It does so rather quickly for most small numbers. All the positive integers less than 10,000 have been tested, and 97.5 per cent of them produced palindromes in 24 or fewer steps. (Question: Why is 24 a critical number?) But the remaining 2.5 per cent failed to produce a palindrome in the first 100 steps; thus the conjecture is probably false. Here is one bit of evidence against the conjecture: The starting number 196 fails to produce a palindrome in 37,303 steps. At that stage the numbers being reversed and added contain more than 15,500 digits.

Results in the palindrome problem depend on the base being used. We normally count with the base 10. Numbers written in diadic notation (base 2), or with any other base, have different palindromic properties. This perhaps weakens the problem in a certain sense. One feels that a property depending on the base of the number system is somewhat less fundamental to number theory than a property inherent in the numbers themselves, regardless of base.

(2.) Starting with any number, add the squares of the digits to form a new number. Repeat the process on the new number, and so on. In a small finite number of steps you will always arrive at either 1 or 4. If 1, the process iterates at 1; if 4, the process "loops" back to 4 in eight steps. Examples of each case are:

86, 100, 1, 1, 1, . . .
36, 25, 29, 85, 89, 145, 42, 20, 4,
    16, 37, 58, 89, 145, 42, 20, 4, . . .

Investigation shows that for all starting numbers $N \leqslant 10^d$, the first step yields a number $< 100d$ down to $N = 300$. Thus the process is rapidly *reductive;* and it turns out that it is necessary to test only numbers less than 166, and not even all of these, because 58 yields the same sequence as 85 (see example), etc. Thus there is no trouble checking that a 1 or a 4 is always attained; the question is, *why* does this happen? Why do all numbers either lead directly to 1 or enter the 4 loop? The problem appears to depend to a lesser extent than the previous one on the base 10. Is there any general theory covering all bases? Can the problem be generalized in other directions? For instance, what if one adds the cubes of the digits instead of the squares?

(3.) Start with any positive integer N. If N is even, divide it by 2 to obtain the next number in the sequence; if N is odd, multiply it by 3 and add 1 to get the next number. Repeat this process. The conjecture is that the sequence thus obtained will always settle down to 4, 2, 1. Illustration:

58, 29, 88, 44, 22, 11, 34, 17, 52, 26, 13,
    40, 20, 10, 5, 16, 8, 4, 2, 1, 4, 2, 1, . . .

The problem is to prove the conjecture; that is, that starting with any N, the sequence will sooner or later hit a power of 2 and therefore taper off to 4, 2, 1. This depends not at all on the base used: a number is either divisible by 2 or not divisible by 2, regardless of base.

Unfortunately the modern computer has robbed the problem of some of its charm. H. S. M. Coxeter, who proposed it in 1970, stated then that it had been checked for all $N \leqslant 500,000$. Thus if the conjecture fails it must do so for a sequence all of whose terms exceed 500,000. However, if the conjecture is true, which seems more likely, a proof will have nothing to do with computers.

For those interested in trying to construct a

proof, we mention that generalization, at least in one direction, is impossible. Divide the evens by 2, as before, but multiply the odds by 5 (instead of 3) and add 1. In this variant, there exist sequences with cycles other than 4, 2, 1. Example:

13, 66, 33, 166, 83, 416, 208, 104, 52, 26, 13, . . .

Any sequence that hits a power of 2 multiplied by 33 or 83 or 13 will enter this loop and stay there.

☐

The prime factors of the sequence of consecutive composite numbers 1802, 1803, . . . , 1810 are given in the first list:

$$1802 = 53 \cdot 17 \cdot 2 \qquad 24 = 3 \cdot 2 \cdot 2 \cdot 2$$
$$1803 = 601 \cdot 3 \qquad 25 = 5 \cdot 5$$
$$1804 = 41 \cdot 11 \cdot 2 \cdot 2 \qquad 26 = 13 \cdot 2$$
$$1805 = 19 \cdot 19 \cdot 5 \qquad 27 = 3 \cdot 3 \cdot 3$$
$$1806 = 43 \cdot 7 \cdot 3 \cdot 2 \qquad 28 = 7 \cdot 2 \cdot 2$$
$$1807 = 139 \cdot 13$$
$$1808 = 113 \cdot 2 \cdot 2 \cdot 2 \cdot 2$$
$$1809 = 67 \cdot 3 \cdot 3 \cdot 3$$
$$1810 = 181 \cdot 5 \cdot 2$$

The largest factor of each is a different prime. This is guaranteed as soon as one knows that the least of these largest prime factors is already greater than the length of the sequence.

The second list carries no such guarantee, because $3 < 5$. But by dipping to 2 as a factor of 24, we can still select a different prime factor from each number of the sequence.

C. A. Grimm has conjectured that this is always possible. That is, in any sequence of consecutive composite integers it is possible to select one prime factor from each without duplication. The conjecture is unproven.

John E. Walstrom and Murray Berg have invented something that they call a *prime prime:* by successively dropping the right-hand digit, the remaining numbers are also prime. Thus 317 qualifies as a prime prime, because 31 and 3 are also primes. Walstrom and Berg have found only twenty-seven primitive prime primes among all numbers of 8 or fewer digits. (31, for example, is not considered primitive because it is already included in 317.) Their search disclosed none at all of 9 digits. Are there no more? If more, how many? Is the supply finite? Very little is known about these peculiar curios.

☐

Suppose we want to determine the decimal expression for $\frac{1}{13}$ by dividing 13 into 1. In the hope that you are agile at mental arithmetic, and to save space, we will use short division. If you prefer to do it by long division, the identical discussion applies. The exact form of a short division varies with the individual arithmetician; one way looks like this:

$$13)1.\ ^10\ ^{10}0\ ^90\ ^{12}0\ ^30\ ^40\ ^10\ ^{10}0\ ^90\ \ldots$$
$$\overline{\ 0\quad 7\quad 6\quad 9\quad 2\quad 3,\quad 0\quad\ 7\quad 6\ \ldots}$$

Thus $\frac{1}{13}$ has a *periodic* decimal expansion, of period 6 units long. The reason the expansion became periodic was that in the sequence of remainders 1, 10, 9, 12, 3, 4, 1, 10, etc., we happened to come around to a 1 again after six steps. Such a repetition need not have occurred at the seventh step, but it *must certainly* have occurred by at most the thirteenth step, because there are only twelve different remainders less than 13. Thus by the nature of the division process, every common fraction $1/n$ has a repeating decimal of period length at most $n - 1$. Once we arrive at a remainder we used before, the whole process repeats from there.

If we call $n - 1$ the maximal period, the question naturally arises, what, if any, integers n have reciprocals $1/n$ whose decimal periods are maximal? The first is 7:

$$7)1.\ ^10\ ^30\ ^20\ ^60\ ^40\ ^50\ ^10\ \ldots$$
$$\overline{.\ 1\quad 4\quad 2\quad 8\quad 5\quad 7,\quad 1\ \ldots}$$

Note that each of the possible remainders, 1 through 6, appears before a repetition occurs.

There is no way known at present of predicting which $1/n$ have maximal periods. The next happens to be 17. A necessary condition is that n be a prime, but this is not sufficient, as we noted when $n = 13$. The maximal n's are scattered in apparently random fashion among the primes. One might expect that they would "thin out" with increasing n—that is, that they would become relatively scarcer, even among the primes; present information is insufficient to decide whether this is the case. About one-third of the first few primes have the maximal property, a proportion which remains approximately constant among the primes up to 1000. I do not know whether it has been checked for larger primes; but if it should turn out that the ratio of maximal n's in any interval to the number of prime n's in that same interval does indeed approach a non-zero constant as N increases, then we have a class of numbers whose asymptotic density is not equal to but proportional to $1/\log N$.

The question has sometimes been raised, how many primes are there of the form $(10^n - 1)/9$? It is easily seen that these numbers consist entirely of a string of ones. To date the only known values of n which give primes are 1, 2, 19, and 23, yielding 1, 11, and the numbers consisting of nineteen and twenty-three 1's respectively. There may be no other n's which give primes, or there may be an infinite number: nothing more is known except a partial list of the composite cases.

The problem is closely connected to the repeating periods of the reciprocals of primes.

□

We mention one more unsolved problem about primes: Is there always at least one prime between every pair of consecutive squares? That is, is there always a prime between, for instance, 100 and 121, and between 625 and 676, etc.? The asymptotic density of the perfect squares is $1/(2\sqrt{N})$, which becomes rapidly smaller than $1/\log N$. The *probability* that two squares bridge a gap in the prime sequence tends to zero as N increases; but that is not what is asked for. As usual, to answer a specific question about the primes is far more difficult than to answer a probabilistic or distributional question about them. And all attempts to link the primes to algebraic entities like perfect squares have so far met with complete failure.

□

The symbol n! means the product of all integers up to and including n. Thus $4! = 1 \times 2 \times 3 \times 4 = 24$. It happens that $n! + 1$ is a perfect square when $n = 4$, 5, or 7. These are believed to be the only values of n with this property; but a tentative proof offered in 1950 was subsequently shown to contain an invalidating oversight. It seems almost as difficult to connect the squares with the factorials as it is to connect them with the primes.

☐

Some numbers, like $13 = 9 + 4$ and $17 = 16 + 1$ can be expressed as the sum of 2 squares. But others, like 7, require 4 squares: $7 = 4 + 1 + 1 + 1$, and there is no sum with fewer squares because none are available except 1 and 4. Fermat proved that it is possible to express every positive integer, however large, as a sum of 4 or fewer squares; no integer requires 5 or more.

Is there a similar greatest number necessary and sufficient for decomposition of cubes, fourth powers, etc.? This is known as Waring's problem, after its originator. Within recent years it has been proven that every number can be expressed as the sum of 9 or fewer cubes. $23 = 8 + 8 + 1 + 1 + 1 + 1 + 1 + 1 + 1$ is a number that actually requires nine.

If negative integers are admissible, the situation is quite different. Mordell says that it is easy to prove that every number can be expressed as the sum of at most 5 integer cubes, positive or negative; and that there is an unproved conjecture that four cubes suffice. For example, 23 now becomes expressible as $8 + 8 + 8 - 1$.

Nineteen fourth powers are required to make up 79, and it is conjectured that 19 is the answer to Waring's problem for fourth powers.

If we let $g(k)$ be the solution of Waring's problem for exponent k, then we have $g(2) = 4$. Furthermore, 4 squares are actually required in an infinite number of cases. For $k = 3$, however, the situation is different. $g(3) = 9$, but only two integers, 23 and 239, actually need 9 cubes. It has further been proved that only a finite number of integers need 8 cubes, and it is conjectured that only a finite number need 7. The greatest integer known to require 7 cubes is 8042. If we call $G(k)$ the largest number of kth powers required for an infinite number of integers, then $G(2) = g(2) = 4$, but $G(3) \neq g(3) = 9$. Perhaps $G(3) = 6$. It is known only that $4 \leqslant G(3) \leqslant 7$. On the other hand, it is known that whereas $g(4)$ is (presumably) 19, $G(4)$ is definitely 16.

$\square$

Euclid defined a perfect number as one which is equal to the sum of all its different divisors, like $1 + 2 + 3 = 6$. The number itself is of course not counted as one of the divisors, or no number could be perfect. Eight was called deficient, because $1 + 2 + 4 < 8$, whereas a number like 12 was said to be abundant: $1 + 2 + 3 + 4 + 6 > 12$. When number theory was still entangled in the meshes of numerology and mythology, the notion of the perfection or imperfection of a number had a more real significance than merely that of a name. The attachment of intrinsic qualities like goodness, malice, and godliness to numbers persists through human history, reaching down closer to the present day than perhaps we would willingly admit. Is it not true that 7 still carries a special aura because of the Sabbath day, and that for many people 13 wears a sinister cloak?

The seventeenth-century amateur mathematician Mersenne, an old friend of Fermat, studied numbers of the form $2^k - 1$, about which not very much is known. A Mersenne number $M = 2^k - 1$ may or may not be prime. For $k = 2, 3, 5, 7$ we get $M = 3, 7, 31, 127$, all primes. But if $k = 4$, $M = 15$, composite; M must always be composite if k is an even number greater than 2, for then

$$M = 2^{2j} - 1 = (2^j + 1)(2^j - 1)$$

In fact, M is composite whenever k is composite; but M need not be prime if k is prime. The first such composite M is $M = 2^{11} - 1 = 23 \times 89$.

The only perfect numbers known to date are of the form $2^{k-1}(2^k - 1)$, where $2^k - 1$ is a *prime* Mersenne number. We lack a test for the primality of Mersenne numbers, which themselves are very large, so that the difficulty of finding perfect numbers is compounded. We do know that every even perfect number is of the above form; but whether odd perfect numbers exist is not known, nor whether there is an infinite or only a finite supply of perfect numbers.

□

The prime factorizations of 220 and 284 are:

$$220 = 2^2 \times 5 \times 11 \qquad 284 = 2^2 \times 71.$$

Now add all the factors of each, prime and composite, including 1 but not the number itself:

220:		284:	
	1		1
	2		2
	4		4
	5		71
	10		<u>142</u>
	11		220
	20		
	22		
	44		
	55		
	<u>110</u>		
	284		

If all the proper factors of a add up to b, and all the proper factors of b add up to a, then a and b are said to be *amicable* numbers. There are many amicable pairs; their number is conjectured to be infinite, but a proof is lacking. There are pairs of odd amicable numbers (the smallest such being 12285 and 14595), but to date no pair has been found with one member odd and the other even. Is such a pair possible? All the known odd amicable pairs contain the factor 3. Is this a necessary condition?

The numbers $12496 \rightarrow 14288 \rightarrow 15472 \rightarrow 14536 \rightarrow 14264 \rightarrow 12496$ form what has been called a *sociable chain* of 5 links, for the obvious reason. The sum of factors of 12496 is 14288, the sum of factors of 14288 is 15472, and so on around the chain. A sociable chain of 28 links had also been known for fifty years, but these were the only two until, in 1970, nine more, all of 4 links, were found by computer. They all consist of large numbers (7 or 8 digits in length). Their discoverer states that his program searched for all chains up to 10 links long in the range of numbers up to 60,000,000 and found only the nine 4-link chains. He conjectured that there may be an infinite number of 4-link chains.

A perfect number could be considered a chain of 1 link, an amicable pair is a chain of 2 links, and now we have also chains of 4 and 5 links. Why do none of 3 links exist? Or do they?

□

Although there are numbers that can be decomposed into the sum of two kth powers in two different ways for $k = 2$, 3, and 4, no such double decomposition is known for any k greater than 4, nor is it known whether such decompositions are possible.

Ever since its publication in 1920, the standard reference work in number theory has been Eugene L. Dickson's great three-volume *History of the Theory of Numbers*. It is a catalog of all published contributions to this large branch of mathematical knowledge. "If it isn't in Dickson, it isn't in number theory." In Chapter 22 of Vol. 2, Dickson discusses the problem of finding an integer decomposable into the sum of two fourth powers in two different ways. The number 17, for example, is so decomposable in only one way:

$$17 = 1^4 + 2^4$$

Several solutions to the problem are known, and Dickson remarks that

$$635318657 = 158^4 + 59^4 = 133^4 + 134^4$$

is said to be the smallest. I was therefore not a little surprised when one of my students announced that he had found a much smaller one. It developed that he was indulging in some mathematical high jinks at my expense, and had extended the problem to include decomposition into *Gaussian* integers, numbers of the form $a + bi$, where a and b are ordinary integers and $i = \sqrt{-1}$. Then

$$82 = 1^4 + 3^4 = (2i - 5)^4 + (2i + 5)^4$$

He further proved that 82 is the smallest natural number decomposable into two essentially different sums of fourth powers. (The other obvious decompositions of 17 such as $i^4 + 2^4$, $i^4 + (2i)^4$, are not essentially different from $1^4 + 2^4$.)

☐

The field of Diophantine equations, those equations requiring solutions in whole numbers only, contains a host of difficult problems which continue to tantalize the number people. It happens that $3^2 + 4^2 = 5^2$ and that $3^3 + 4^3 + 5^3 = 6^3$. But this is just luck. It is not true that $3^4 + 4^4 + 5^4 + 6^4 = 7^4$. This gives rise to the question of whether there are *any* other integers a, k, and m satisfying

$$a^m + (a + 1)^m + \cdots + (a + k)^m = (a + k + 1)^m$$

The problem has been extended further. The system of equations:

$$a^3 + b^3 + c^3 + d^3 = x + y + z$$
$$a^6 + b^6 + c^6 + d^6 = x^2 + y^2 + z^2$$
$$a^3 + b^3 + c^3 = d^3$$

has solutions $a = 3$, $b = 4$, $c = 5$, $d = 6$, $x = 91$, $y = 152$, $z = 189$. Find the general solution in integers, or prove that this is the only one.

Another problem: Find all possible pairs

(m, n) such that the sum of the first m consecutive integers equals the sum of the first n consecutive squares. For example, $(m, n) = (10, 5)$ satisfy:

$$1 + 2 + 3 + 4 + 5 + 6 + 7 + 8 + 9 + 10$$
$$= 1^2 + 2^2 + 3^2 + 4^2 + 5^2$$

Two other pairs are known to satisfy: $(13, 6)$ and $(645, 85)$. Are there any more? An infinite number?

The formula for the sum of the first m integers is $m(m + 1)/2$. The sum of the first n squares is $n(n + 1)(2n + 1)/6$. Therefore the problem is to find the general solution of the Diophantine equation

$$3m(m + 1) = n(n + 1)(2n + 1)$$

□

Some typical unsolved Diophantine problems follow.

(1) Is there an infinite number of primes of the form $n^2 + 1$? The first few are given by $n = 1, 2, 4, 6, 10, 14, 16, 20$. The same question can be asked for numbers of the form $p^2 + 1, p$ a prime. If the answer to either of the above questions is yes, then what is the asymptotic density of (a) the set of numbers of the prescribed form, and (b) the set of generating n's or p's?

(2) Are there three integers whose product

equals the cube of their sum? In other words, does the Diophantine equation

$$(x + y + z)^3 = xyz$$

have a solution? Don't forget that negative integers are admissible. This looks like an easy problem; yet it has been unsuccessfully begging to be answered for some time.

(3) It is known that the number of integral solutions of

$$x^3 - y^2 = 7$$

is finite, but we do not know all the solutions, nor even how many there are.

(4) Are there any two consecutive numbers of the form a^b, both a and b positive integers, except 2^3 and 3^2?

(5) Is there an infinite number of triples of *consecutive* primes in arithmetic progression, like 3, 5, 7 and 47, 53, 59?

(6) For what m does the equation

$$x^3 + y^3 + z^3 + w^3 = m$$

have (positive or negative) integer solutions in x, y, z, w? This is another form of the conjecture mentioned on page 109. What is wanted is a test, valid for all m.

(7) Even when it is known that a certain m gives a solution of (6), it is not known whether

there exists an infinite number of other solutions for that m.

(8) The conjecture of (6) can be strengthened by asking whether there is always a solution even when we require that $w = z$. That is, does

$$x^3 + y^3 + 2z^3 = m$$

have a solution for every m? The least value of m for which no solution is known, nor even whether a solution exists, is 76.

(9) The sharper requirement

$$x^3 + y^3 + z^3 = m$$

is known to have solutions for some m, but there are many unanswered questions. For instance:

(a) Does it have only two solutions for $m = 3$? (They are $1, 1, 1$ and $4, 4, -5$.)
(b) Does it have any solution for $m = 30$?
(c) Is there any value of m for which it has infinitely many solutions?

(10) What are the conditions on general a, b, for solvability in integers of the equation $x^2 + y^2 + z^2 - axyz = b$?

(11) Does the Diophanatine equation $ax + by = c$, with a and b relatively prime (no common factor except 1), have an infinite number of solutions such that x and y are both primes?

(12) If the answer to (11) is yes, is it true in the special case $a = 1$, $b = -1$, $c = 2$? This is the twin prime conjecture.

(13) The equation $x^4 + y^4 + 64 = z^4$ has

the solution-triples (x, y, z): $(1, 2, 3)$; $(7, 8, 9)$; $(21, 36, 37)$. Are there any others? An infinite number? Are there any with $z - y \neq 1$?

(14) Conjecture: If p is a prime congruent to 3 modulo 4 (meaning it has remainder 3 when divided by 4), and if U is the value of x in the least primitive solution of the Diophantine equation $y^2 = px^2 + 1$, then U is never evenly divisible by p. Examples:

$$4 = 3 \times 1^2 + 1$$
$$64 = 7 \times 3^2 + 1$$

The conjecture has been verified for all $p < 18{,}000$, but never proved.

The list could go on almost indefinitely. If you are interested in Diophantine analysis, the field, like most difficult ones, is wide open.

□

Let $f(x, y)$ be a general cubic equation in x and y with rational coefficients. No general method is known for finding a solution of this equation in rational numbers—that is, a point on the curve, both of whose co-ordinates are rational.

H. Davenport has proven that a homogeneous cubic equation in n variables $f(x_1, x_2, \cdots x_n) = 0$ always has solutions in integers (not all zero) provided $n \geqslant 16$; but this is probably not best possible. Davenport guesses that 10 might be.

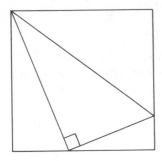

[40]

Does there exist a rectangular parallelepiped (box) all of whose edges and face diagonals are of integral length, and the length of whose main diagonal is also an integer? This is a Diophantine problem, a sort of extended Pythagorean theorem.

Another one is, can a right triangle be inscribed in a square in the manner shown in Figure 40, such that the length of every line segment in the diagram is an integer?

☐

Single, isolated number questions are often difficult to attack because they fail to fall into any known pattern or category. Possibly for this reason also they possess less charm than some of the bigger problems. One such numerical question is to determine whether or not $2^7 = 128$ is the only power of 2 of two or more digits each of which is a power of 2 (recalling that $1 = 2^0$). Professor R. J. Walker, who proposed the problem, supplies a hint for those interested in attacking it. It is not possible to find a positive integer, r, large enough so that no power of 2 could have its last r

digits all powers of 2. The opposite is true. In fact, for any integer r there exist powers of 2 such that all of the last r digits are either 1 or 2.

A positive number each of whose prime factors occurs at least twice in the factorization has been called a powerful number. Thus 100 is powerful, not because 10^2 appears in its factorization, but because its *prime* factors all appear twice: $2 \times 2 \times 5 \times 5$. Whereas $72 = 2^3 \times 3^2$ is powerful, $12 = 2^2 \times 3$ is not.

The only known instances where the difference between two non-square powerful numbers is less than 4 are: $2^3 \times 3^2 \times 13^2 - 23^3 = 1$, and $2^7 - 5^3 = 3$. The smallest such difference known to occur infinitely often is 4.

It is an unproven conjecture that 6 is never the difference between any two powerful numbers, square or non-square. It may be that there are infinitely many other such impossible differences —or none.

□

Note that, through the usual process of rationalizing the denominator,

$$\frac{1}{\sqrt{2}+1} = \frac{1}{\sqrt{2}+1} \cdot \frac{\sqrt{2}-1}{\sqrt{2}-1} = \frac{\sqrt{2}-1}{2-1}$$
$$= \sqrt{2} - 1.$$

From this we get

$$\sqrt{2} = 1 + \frac{1}{1+\sqrt{2}}.$$

Now if we replace the square root of 2 that appears on the right by the whole expression for $\sqrt{2}$, we have

$$\sqrt{2} = 1 + \cfrac{1}{1 + 1 + \cfrac{1}{1 + \sqrt{2}}}$$

$$= 1 + \cfrac{1}{2 + \cfrac{1}{1 + \sqrt{2}}}.$$

Repeating this substitution yields what is known as the continued fraction expansion for $\sqrt{2}$:

$$\sqrt{2} = 1 + \cfrac{1}{2 + \cfrac{1}{2 + \cfrac{1}{2 + \cfrac{1}{2 + \cdots}}}}.$$

We know that this process is valid if we stop at any specific stage and leave a final $\sqrt{2}$ on the right. Whether it has any meaning as an *infinite* continued fraction depends on the question of convergence. It is not difficult to show, fortunately, that continued fractions of this type do converge.

It turns out that each irrational number has a *periodic* continued fraction expansion. The period length, L, for $\sqrt{2}$ is 1. For $\sqrt{3}$ we have

$$\sqrt{3} = 1 + \cfrac{1}{1 + \cfrac{1}{2 + \cfrac{1}{1 + \cfrac{1}{2 + \cdots}}}}$$

which has period length $L = 2$.

The question of period length of the repeating decimal of a rational fraction has an analog here in the period of the repeating part of the continued fraction expansion of a quadratic irrational. How much is known about these periods? Very little. In one of the few available books on the subject (see Notes), there is a table of the periodic part of the expansion of $\sqrt{N}$ for all N up to 40. Is 8 the longest possible period? (31 has period of length 8.) Why are there no periods of length 3? These two questions, as it happens, have a very simple answer: the table isn't big enough. $\sqrt{43}$ has $L = 10$, and $\sqrt{41}$ has $L = 3$. One would like to have the answer to more general questions, which undoubtedly require considerable investigation. Is there any maximal period length, or is there no upper bound on L? Is there a formula for L in terms of N? If that is too much to ask, is there any way at all of predicting L? What is there about two different values of N that give them the same L? Is every positive integer an eligible candidate for some

L? Almost no work appears to have been done on this subject.

☐

Starting with $a_0 = 1$ and $a_1 = 1$, let the law of formation of subsequent members of a sequence be $a_{n+1} = a_n + a_{n-1}$. This law generates the famous Fibonacci Sequence:

1, 1, 2, 3, 5, 8, 13, 21, 34, 55, 89, 144, 233, . . .

Many interesting relations come out of a study of the Fibonacci numbers.

In how many ways can a number be represented as a sum of different Fibonacci numbers? When a number theorist of the stature of Leonard Carlitz remarks, "Unfortunately the general case is very complicated," we can be sure that the problem will not succumb to a casual attack.

The third, fifth, seventh, and eleventh numbers of the sequence are primes. We are thus tempted to guess that, if F_n stands for the *n*th Fibonacci number, F_n is prime whenever *n* is a prime >2. But this pretty conjecture fails at an early stage: F_{19} is composite. Not only is there no known device predicting which F_n are prime, but it is not even known whether the number of prime Fibonacci numbers is finite. If it is infinite, what is their asymptotic density?

It is not difficult to show that every integer

is a *periodic factor* of the Fibonacci numbers. The number 3, for example, is a factor of every 4th member of the sequence, and thus has period 4 in this sense; 5 has period 5; 11 has period 9. If p is a prime, is it possible that p and p^2 have the same period? This does not seem likely; but the answer to the question is unknown. In fact, as D. M. Bloom aptly suggests, "It seems that the Fibonacci numbers still provide much material for investigation."

6

Topological problems

Numbers are old friends: we grew up with them and we see them every day. They have become part of our life. Although numbers are really very abstract entities, we nevertheless find them easy to understand through long familiarity and training.

The field of topology is quite another story. We cope with it every day too, but without thinking and without systematizing our language and our techniques as we have done with numbers. We know that a left glove will not fit a right hand; yet why does a telephone receiver, shaped to curve between mouth and ear, fit either side of the face? Why do some configurations of a piece of string constitute knots whereas others are mere loops that disappear when the string is pulled? How are the

[41]

three rings in the trade mark of a well-known brewery linked together, or are they linked at all? (See Figure 41: no one ring passes through any other ring; yet they cannot be pulled apart.) That every direction is south to a person standing at the north pole is the fault of our latitude-longitude system. Inasmuch as the polar regions are becoming more important places than they once were, it might be convenient to adopt a co-ordinate grid on which there would be no such difficulty anywhere on earth. Is such a grid possible?

These are topological questions. Their formulation and consideration by mathematical methods was not even attempted until recent times. Topology is a twentieth-century product; hence one might expect that it abounds with unsolved problems, and it does. But although some of them are easily stated, for most of them we require first a language in

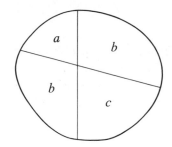

which to frame the question. This language, unlike that of the number system, is not familiar to most of us.

The oldest and best-known unsolved problem in topology is also the easiest to explain. It is the famous four-color problem. In making a map, one wishes to use different colors for any two countries with all or part of a border in common. If two countries come together at only one point, like *a* and *c* in Figure 42, they are not considered to have a common border; for they could be colored alike without confusion. Some maps require four colors, as in Figure 43. If this map represents an island, the ocean can either be disregarded or colored the same as *b*: in either case no more than four colors are required, but three will not do.

The question is, are there any maps which require five colors? None are known, but on the other hand no proof exists that four colors are always sufficient. (It is known that *five*

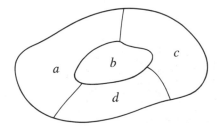

colors are always sufficient.) The problem has been driven out to this stage: it has been shown that no map of fewer than thirty-five regions can require more than four colors. Thus if any five-color maps exist, they are very complicated ones.

The coloring problem is the same on the surface of a sphere as it is on a plane, but on the surface of a torus (doughnut) it is quite different. If the earth were a torus it might be necessary to use seven colors to map its surface, but surely no more than seven. Oddly enough, this seemingly more difficult theorem has been proved.

What is the least number of colors with which one can color a plane map in such a way that no two points one unit apart are ever the same color? That seven colors suffice is demonstrated in Figure 44. The pattern can be extended. Each number represents a different color, and the diameter of each hexagon is slightly less than 1, say 0.99. The question is, are seven colors necessary, and if not, what is the minimum number?

☐

[44]

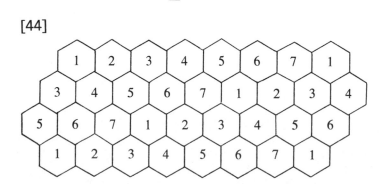

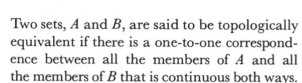

45

Two sets, *A* and *B*, are said to be topologically equivalent if there is a one-to-one correspondence between all the members of *A* and all the members of *B* that is continuous both ways.

One-to-one means just what it says: to every member of *A* can be matched one member of *B*, no more and no less, and vice versa. Note that we do not have to *count* the members of either set to be able to decide whether such a correspondence exists. The phrase means only that such a correspondence *can be* observed. Thus if *A* is the set of points on a line segment 1 inch long and *B* is the set of points on a line segment 2 inches long, then *A* and *B* are topologically equivalent. One need only demonstrate that they can be matched one-to-one, as shown in Figure 45. In precisely the same way, a semicircular arc 1 inch long with the end points of the semicircle deleted is equivalent to a straight line of infinite length. In Figure 46, each radial line matches one point of the arc with one point

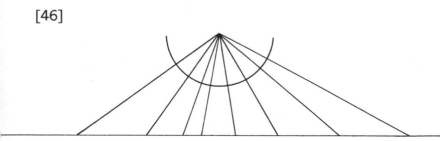

of the line. The "continuity both ways" requires, loosely speaking, that points *close to* point P of one set are matched with points close to point P', the image or map of P, in the second set. The definition of continuity can be made precise, but this one will do for our purposes.

A one-to-one correspondence of the type described above is called a homeomorphism, and two sets are topologically equivalent if they are homeomorphic.

One might suppose that a homeomorphism is equivalent to a deformation transformation, but that is only one kind of homeomorphism. Two sets have been described as topologically equivalent if one can be transformed into the other by "kneading and stretching without breaking and tearing." This is true, but it is not the whole truth. The set of all points on two spherical surfaces (of different radii) tangent internally is topologically equivalent to the set of all points on the same two surfaces tangent externally: the necessary homeomorphism exists. But it is not possible to transform the first set into the second by a pure deformation.

Consider now a circular disk: the set of all points on and inside a circle in the plane. An interesting theorem of L. E. J. Brouwer states that any homeomorphism of such a disk onto itself leaves at least one point fixed. Two questions connected with this transformation are raised by Ulam. If a homeomorphism maps the points p of a disk onto the points p' of the disk, then do there exist arbitrarily small triangles $p_1p_2p_3$ congruent to the triangles formed by connecting the image points $p'_1p'_2p'_3$? One might think that such triples of points could be found in the neighborhood of the fixed point. On the other hand, some such triples may occur in an entirely different part of the disk. Note that the question does not demand that any one of p_1, p_2, p_3 remain fixed, nor does it say that a whole triangle must be lifted bodily by the homeomorphism and set down in a new position. It asks only about some three distinct points. If the answer to the above question is yes, a second question is, do such triangles exist with prescribed angles?

"Satisfactory necessary and sufficient conditions have not been found for determining whether or not a set has the fixed point property. The sufficient conditions that have been found are too restrictive to be necessary. On the other hand, many examples have been shown to have the fixed point property—sometimes with the method of proof tailored to the example."

□

Two curves in ordinary three-dimensional space are said to be mutually *enlaced* if there exists no homeomorphism of the whole space under which the images of the two curves are contained in disjoint geometric spheres. There is at present no workable analytical criterion by which one can decide whether two given curves are enlaced.

A topological property somewhat different from enlacement is *knottedness*. The characterization of knots is by no means complete. The following discussion goes beyond the knot itself to the 3-space surrounding it. "It would be interesting to describe the system of magnetic lines of force due to a current flowing on a knotted (infinitely thin) wire. In particular, suppose the current flows through a 'cloverleaf' knot. Does the system of lines of magnetic force in space surrounding the knot reflect topologically the knottedness of the curve? Such systems of curves may exhibit considerable topological complexity even when generated by currents flowing on straight lines, as shown by calculations on the properties of lines of force due to currents flowing on the three straight lines $x = 1, y = 0; y = 1, z = 0; z = 1, y = 1$."

7

Probability
and combinatorial problems

There is no branch of mathematics quite so deceptively tricky as probability theory. Doubtless more erroneous answers have appeared in print to questions in probability than in any other field. Even the experts have occasionally been deceived. Let these words serve as a warning to any novice too eager to plunge into a problem that asks, "What are the chances . . . ?"

To illustrate one of the pitfalls we cite a famous easy example. There are three cards, with no markings on them, alike except for their coloring. One is red on both sides, one is red on one side and white on the other, and the third is white on both sides. After shuffling them in a closed bag we draw one out and lay it on the table, no one having had any opportunity to see the side that is down. Suppose the side showing is red. Then obviously it is not the white-white card, and it must be one of the other two. If it is the red-red card the other side is red, and if it is the red-white card the other side is white. It seems as if one might bet even money on either possibility.

But that analysis is faulty. The chances that the under side is red are not even, but two to one in favor. The point, and it is the important point missed in many probability questions, is that the events in doubt must be *equally likely*. There are two possibilities, to be sure; but they are not equally likely. We may be looking at side No. 1 of the red-red card, or at side No. 2 of the red-red card, or at the red side of the red-white card. These three possibilities *are* equally likely. Two of them will lead to red on the concealed face, only one to white.

☐

A perfectly flexible piece of string (or theoretical string) of length s is thrown down on the floor "at random." If the distance between the two ends is then carefully measured, what is the average, or *expected value,* of this distance measurement? The question sounds as if it makes sense, but it may well be meaningless. The difficulty here is more subtle than that of the 3-card problem. It is well illustrated by the following better-known example.

Suppose we ask someone to draw "at random" a chord in a given circle. What is the probability that this chord will be longer than the side of an equilateral triangle inscribed in the circle? There are at least three ways of defining the randomness of the chord. (1) We can select a point A on the circle and consider all chords with one end at A. If the other end of a chord is equally likely to lie on any other point of the circle, we can distribute points uniformly around the circle to serve as these other ends. Now inscribe the triangle with a vertex at A (Figure 47a). It appears equally likely that the random chord should intersect one of the three equal arcs AB, BC, CA. Thus the probability that it is longer than the line AB is ⅓. (2) If we distribute our sample chords at equally spaced intervals along a diameter CD (Figure 47b), we get the result that half of them are now longer than AB, so that the answer to the probability question is ½. (3) A third possibility is to consider a chord to be determined by the position of its midpoint. Note that each point inside the circle (except the

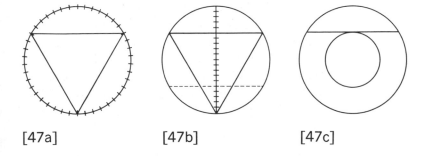

[47a]   [47b]   [47c]

center) can serve as the midpoint of precisely one chord. If r is the radius of the given circle, all midpoints located inside the concentric circle of radius $r/2$ will have chords longer than AB (Figure 47c). The ratio of such points to the total number of available points appears to be proportional to the areas involved. The area of the inner circle is $\pi r^2/4$ and of the outer one πr^2, so that the required probability is now $\frac{1}{4}$.

These three probabilities are drastically different. They cannot all be "right." If we ask, what is the average, or expected length of a random chord, we cannot anticipate a meaningful answer without some further definition of "random." The three interpretations will each yield a different expected length. If we select a point on the circumference, pivot a thin rod at that point, spin it, and draw a chord in the direction at which it comes to rest, we have an operational procedure for randomizing in the sense of (1). If we observe the trajectories of a great many stars passing across the field of view of a telescope turning with the earth and aimed at the sky's equatorial region, we will have chords of type (2). Suppose we draw a large number of equal non-overlapping circles on a smooth floor and then drop a needle on the floor. If the point of the needle does not fall within any circle, we disregard that throw. If it falls within a circle, we let that point be the midpoint of a chord. A large number of such throws should yield chords randomized according to (3). Many equal circles are used rather than just one to minimize the "target effect" of aiming the needle at something.

The notions of *random* and *equally probable* are not adequately defined in this problem, so that although we may think we know what a random chord is, in fact we do not.

In the problem of the thrown string, the same difficulty is encountered. No one has been able to suggest any satisfactory mechanical method for randomizing the throws, and without some further definition, we do not know what a random throw means. In guessing at a solution one might reason as follows. If the string is of length s, and the distance between the ends after throwing is d, then the greatest possible value of $d = s$, and the least possible value of $d = 0$, and hence the average ought to be $s/2$. But when the problem appeared in print, several theoretical "solutions" were submitted, all staunchly supported by their proponents—and all different. Yet they seemed all to favor values somewhere near $s/3$ rather than $s/2$. Also a few inconclusive experiments tended in the direction of $s/3$. Nevertheless the originator of the problem still inclines to the view that the problem is too ill-defined to make any solution acceptable.

□

The study of probability hinges on an analysis of the permutations and combinations involved. That some combinatorial problems are quite easy to state but very difficult to solve is evidenced by the postage stamp problem: in how many ways is it possible to fold a strip of n postage stamps? The folds are to occur only along the perforations between stamps, and a strip means the kind that comes out of a coil machine, one stamp wide. Of course the answer for certain small values of n can be found by actual experiment. What is wanted is a formula or an analysis for arbitrary n.

A generalization with an added dimension of difficulty is: given a road map of N folds, in how many ways can it be re-folded? Faced with a stuffed glove compartment, you have probably already worked with this problem!

Starting from the origin, take a "step" one unit long in any of the four cardinal directions, up, down, right, or left, so that this step ends at one of the four nearest lattice points. Then take a second unit step, "at random" to the extent that is equally likely to be in any one of the four possible directions, and continue in this fashion n times. Such a sequence is called a *random walk of* n *steps*. One random walk of 15 steps is shown in Figure 48. Note that this particular walk is special in that it never crosses itself; more than that, it never touches the same lattice point twice. How many random walks of this type are there for each n? As

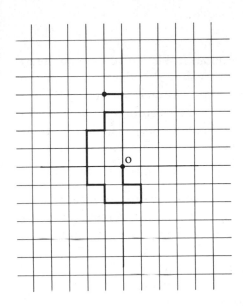

[48]

usual, it would be nice to have the answer expressed as a function of n. It will, of course, be something less than 4^n.

□

What is the probability that when N points are chosen at random in a unit cube, no two of them will be closer to each other than a prescribed distance, δ? "At random" here means that all points are equally probable. "This is an important unsolved problem in statistical mechanics. Its solution would shed much light on the theory of changes of state of matter (for example, from solid to liquid, as in melting)."

What is the expected value of the volume of a tetrahedron whose vertices are four points chosen at random inside a given regular tetrahedron of unit volume? Is the answer different if the given tetrahedron is not regular?

Here is an abstract combinatorial problem: The distinct pairs of n objects, n odd, are arranged in n columns so that each column contains $\frac{n-1}{2}$ pairs and no one object occurs more than once in the same column, without regard to the order of the column or of the pairs in a column. Determine the number of different ways in which this can be done. To illustrate, if $n = 3$ and the objects are A, B, and C, then the only pairings are AB, AC, and BC, and since the columns are only one pair long, that is the total answer: one way. If $n = 5$, A, B, C, D, and E can be paired and placed in columns in several ways. Two of them are:

$$(1) \quad \begin{array}{ccccc} \text{AB} & \text{AC} & \text{AD} & \text{AE} & \text{CE} \\ \text{CD} & \text{DE} & \text{BE} & \text{BC} & \text{BD} \end{array}$$

$$(2) \quad \begin{array}{ccccc} \text{AB} & \text{AC} & \text{CD} & \text{AE} & \text{CE} \\ \text{DE} & \text{BD} & \text{BE} & \text{BC} & \text{DA} \end{array}$$

How many ways are there? More generally, how many ways when 5 is extended to n?

☐

Hubert Phillips, who conducted a famous problem column under the pen name Caliban, once posed the following.

"*The five red balls.* The Professor had a number of balls of various colours. He put a certain number (the colours of which were not known) in a bag, and the class then drew out five at random. All five were red. 'That that would happen,' said the Professor, 'was exactly an even money chance.' How many balls in all had he put in the bag, and how many of them were red ones?"

Caliban's published solution was that the bag originally contained nine red balls and one not-red ball. Some further questions suggest themselves. (1) Is the Caliban solution the only one possible for five? (2) Can the problem be solved if "five" is replaced by "six" throughout? (3) Is there a general solution? Because the mathematics becomes technical, the interested reader is referred to the notes for a discussion of these questions.

□

Each of n points in the plane is joined to every other one by continuous curves. What is the least number of intersections of these curves?

If this least number is called X_n, we note from Figure 49 that four points can be mutu-

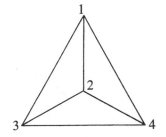

[49]

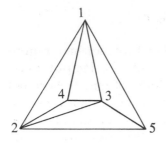

ally interconnected with no intersections, so that $X_4 = 0$. Figure 50 seems to indicate that points 4 and 5 cannot be connected without an intersection with a previously drawn line. This has in fact been proven, so that $X_5 = 1$. It is probably true that $X_6 = 3$, but nothing further is known. What is wanted is a general formula for X_n or a procedure for finding X_n. The same problem on the surface of a sphere is also essentially unsolved.

Paul Erdös has conjectured that every plane convex polygon of n vertices contains a vertex such that the number of different distances from it to the other vertices is at least $n/2$ if n is even, and $(n - 1)/2$ if n is odd.

☐

Inscribe in (different) circles, first a straight line, then a triangle, then a quadrilateral with all its diagonals, then a pentagon with all its diagonals, and so on. The inscribed n-gons need not be regular. In fact, they must not be regular when n is an even number exceeding 4, to ensure against multiple crossings of the diagonals at a single point. Into how many regions is the circle partitioned in each case? More precisely, what is the maximum number of regions into which a circle is partitioned by an inscribed n-gon and all its diagonals?

By counting the areas in Figure 51, we find 2, 4, 8, 16 regions respectively and perhaps conclude that the answer is 2^{n-1}. If so, Figure 52 should have 32 regions. We count and find 31. Must have missed one; count again: still 31. In fact, 2^{n-1} is wrong, and is another example of jumping at an unwarranted conclusion. In mathematics, four swallows do not make a summer, nor do four cases establish a general formula.

For the 7-gon we have not 64 nor even 63,

[51]

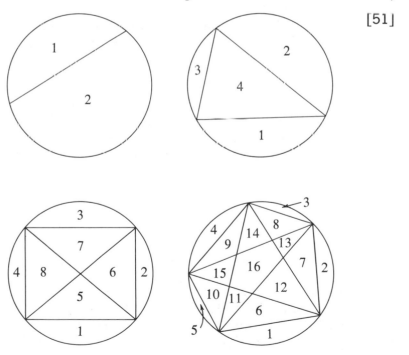

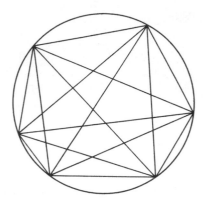

but 57 regions. The correct formula is known to be

$$n + \frac{(n-1)(n-2)(n^2 - 3n + 12)}{24}.$$

The Notes tell where a proof can be found, done in two ways, each of which takes nearly two pages.

In 1970 Nancy Cable, then a sophomore at Skidmore College, observed that the same sequence,

$$2, 4, 8, 16, 31, 57, 99, \dots$$

may be obtained by adding the numbers in each horizontal row to the left of the diagonal line of Figure 53. The interesting question is, why? Can we count these regions using the binomial coefficients in some natural way?

[53]

```
                    1    1
                 1    2    1
              1    3    3    1
           1    4    6    4    1
        1    5   10   10    5  / 1
     1    6   15   20   15  / 6    1
  1    7   21   35   35  / 21   7    1
1    8   28   56   70  / 56   28    8    1
1  9   36   84  126 / 126  84   36    9    1
•    •    •    •    •  /  •    •    •    •    •    •
```

8

A glimpse of some problems
of analysis

A rational number, p/q, is a number expressible as the quotient of two integers, p and q. If the rationals are interpreted as points on the real line, one can always find a new rational point between any two given rationals, however close: their average, for instance. That is to say, the rationals are *dense* on the real line. In the same way, the points whose co-ordinates are both rational are dense in the plane.

Ulam asks whether it is possible to characterize or describe the sets of points in the plane such that the distance between any two of their points is rational. He asks further whether such a set can be dense in the plane. All the rational points on a straight line comprise a set that answers the first question in the affirmative; but a one-dimensional set is not what is wanted.

☐

Questions asking what shape, what path, or what form will yield a minimal or optimal result are known as variational problems. We have already met some in Chapters 1 and 2. A classical question known as Plateau's problem leads to a number of similar questions. J. A. F. Plateau asked for the surface of smallest area bounded by a given closed non-plane curve in space. The general Plateau problem was solved by Tibor Rado and by Jesse Douglas in 1930–31.

What is the shortest curve joining two points on an ellipse and at the same time dividing the area of the ellipse into two equal simply connected pieces? Is the problem solvable if the two points coincide? If a simple closed curve lies on the surface of a sphere, what is the surface of minimal area through this curve dividing the sphere into two equal volumes? Inasmuch as this is Plateau's problem with an added condition, it may not be solvable.

Given a plane convex quadrilateral, find the shortest curve that divides it into two equal areas. The problem has been (partially) solved in two dimensions: the curve must be a circular arc meeting two sides of the quadrilateral

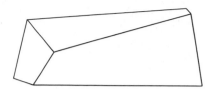

orthogonally. *Which* two sides, and exactly where, depends on the shape of the quadrilateral and would seem to be difficult to specify. In three dimensions the problem is apparently wide open: given an irregular convex "box" with six plane faces (not, of course, a rectangular box but something perhaps like Figure 54, with no two faces parallel or congruent,) find or characterize the minimal surface that divides it into two equal volumes. One should first ascertain whether such a minimal surface exists; it might exist for some but not all shapes.

☐

In *What is Mathematics?*, Richard Courant and Herbert Robbins describe a mechanical method for "solving" various aspects of Plateau's problem. If a wire framework is dipped into soapsuds and carefully withdrawn, a film of soap will cling to the frame, stretched to form a minimal surface because of the surface tension of the film, in the same way that a soap bubble takes the form of a sphere because in that way it contains a given volume of air with the least surface. If a wire model of a cube is dipped into the suds, the result is a system of

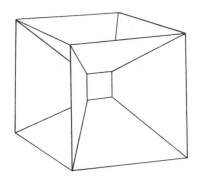

thirteen nearly plane surfaces that take the form indicated in Figure 55. But the surfaces are not all plane, nor is the small central one exactly square. Courant and Robbins call it "a challenging unsolved problem" to find or characterize these surfaces in some analytic fashion. As far as I know, no progress has been made on the problem in the thirty years since the book was published.

□

"Suppose two segments are given in the plane, each of length one. One is asked to move the first segment continuously, without changing its length, to make it coincide at the end of the motion with the second given interval in such a way that the sum of the lengths of the two paths described by the end points should be a minimum. What is the general rule for this minimum motion? . . . One could require alternately [he means alternatively] that instead of the sum, the square root of the sum of the squares of the lengths described by the end-points should be minimum.

"More generally, one could pose an analogous problem of the 'most economical' motion given a geometrical object A and another B congruent to it and requiring the motion from A to B to be such that a sum or integral of the lengths of paths described by individual points be minimum. . . . One motivation behind the consideration of such questions is that in certain problems of mechanics of continua, e.g. in hydrodynamics, the motions that are most prevalent are singled out by extremal principles not unlike the above; but of course operating in a space of infinitely many dimensions."

□

Paul Erdös has proposed (and solved) many nasty questions involving inequalities. Here are three unsolved ones.

(1) If m and n are integers satisfying

$$\left(1 - \frac{1}{m}\right)^n > \frac{1}{2}, \qquad \left(1 - \frac{1}{m-1}\right)^n < \frac{1}{2}$$

prove the relations

$$(m-1)^n > (m-2)^n + (m-3)^n + \cdots + 1^n$$
$$(m+1)^n < m^n + (m-1)^n + \cdots + 1^n$$

Show also that the inequality

$$m^n > (m-1)^n + (m-2)^n + \cdots + 1^n$$

is true in infinitely many instances, but is also untrue in infinitely many instances.

(2) Let $a_1 < a_2 < \cdots < a_n \leqslant 2n$ be a sequence of positive integers. Then

$$\max (a_i, a_j) > \frac{38n}{147} - c$$

where c is independent of n, and (a_i, a_j) denotes the greatest common divisor of a_i and a_j. We are asked to (a) prove the statement, (b) find c, and (c) show that this is a best possible estimate.

(3) Let $a_1 < a_2 < \cdots < a_k \leqslant n$; $b_1 < b_2 < \cdots < b_l \leqslant n$, be two sequences of integers such that all the products $a_i b_j$ are different. Prove that

$$kl < c(n^2/\log n)$$

Erdös says that if this is true, for some universal constant c, then it is a best possible estimate.

□

It is known that there exist infinitely many rational integral algebraic equations with integer coefficients, the leading one of which is unity, with all but one root occurring within a specified interval. If you wished to make use of this fact, it would doubtless be convenient to select one with small coefficients. Is there any procedure by which this can be done? In particular, what is the equation of degree n

fulfilling the above conditions and having in some sense the "smallest" coefficients?

☐

A transcendental number is the solution of no algebraic equation. π and e are transcendental numbers; but it by no means follows that their sum $\pi + e$ is transcendental, although of course it may be. In fact no one has yet been able to prove even that $\pi + e$ is irrational.

The famous equation connecting π and e, $e^{\pi i} = -1$, is not algebraic. A difficult question is whether π and e are connected by any algebraic expression whatever.

Notes

The number refers to the corresponding page of the text

1. A word of apology for the title is in order. Not by any stretch of the imagination could this book be considered a summary of those topics that will be the major concern of tomorrow's mathematicians. Because of the limitations imposed by our efforts to keep the language non-technical, we have been unable to mention many of the best problems that crowd the working mathematician's docket. Furthermore, it is the rash soothsayer indeed who ventures to predict the next turn of events in this ever-changing and expanding science.

7. In the following discussion we describe, with the aid of an ingenious example due to Schwarz, one of the difficulties involved in the attempt to define surface area of a curved surface.

How could you measure the length of a curved piece of string? This is not so difficult as finding the area of a piece of orange peel. The string can be straightened out, and measured with a yardstick; the orange peel cannot be flattened out. However, if we talk about a mathematical curve instead of a piece of string, we are neither permitted nor able to pick it up and straighten it out. What we can do is to find out its length by a *limiting process.* Perhaps you recall from your plane geometry days that that was the way you found, or rather *defined,* the perimeter of a circle. The method is first to inscribe a regular hexagon in the circle, and show how its perimeter can be precisely calculated. Then one inscribes a regular twelve-sided polygon, and again calculates its perimeter. Polygons of 24, 48, 96, . . . sides are then considered, and a method is described to show how their perimeters can continue to be calculated if the number of sides is repeatedly doubled. The next stage is the observation that the perimeters of the polygons *approach* the perimeter of the circle as the number of sides is

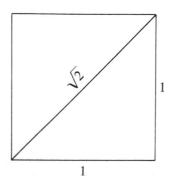

increased. By this process one has not succeeded in writing down the exact length of the
circumference; but two important things have
been accomplished: (1) the length of the circumference has been *defined,* as a limit of known
ascertainable lengths; and (2) the numerical
value of the circumference's length can be
found approximately, and that approximation
can be made as close as we please by carrying
the work to any specified number of decimal
places.

In defining something in terms of a limit,
one must be very careful. Suppose, for example, that we are interested in finding the length
of the diagonal of a square one unit on a side
(Figure 56). The Pythagorean theorem tells us
that the square of the hypotenuse equals the
sum of the squares of the other two sides, so
that what we seek is of length $\sqrt{2}$. But $\sqrt{2}$,
being irrational, is a somewhat elusive creature;
and we might seek to evaluate it as follows.
Draw the diagonal as if it were a staircase with
a large number of very small equally spaced
steps (Figure 57). Then keep doubling and

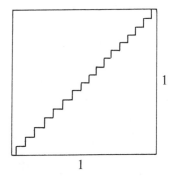

1

1

[57]

redoubling the number of steps. The step-diagonal may *look* as if it were approaching the actual diagonal as a limit. Indeed, we can bring any part of the stepped line to within as small a distance as we please of the original diagonal. But it is easy to see that it won't do as a measure of the *length* of the diagonal; for the length of the stepped line is always the total length of the horizontal elements (the treads) plus the total length of the vertical elements (the risers), or $1 + 1 = 2$, not $\sqrt{2}$.

"But that's a silly illustration," you say. "To guard against such an error we have only to specify that *both ends* of the broken lines that we use in the limit process lie on the original curve (or line). That was how we did the circle, and it worked: the vertices of the inscribed polygons were always points on the circle." And you are quite right. The length of a "smooth" curve can be defined as the limit of the sum of straight-line segments, each of whose lengths is made smaller in a definite specified way so that both ends of all of them always lie on the curve.

Why can't this concept be extended in the natural way to measuring surface areas? For many years mathematicians thought that it could be. Instead of straight-line segments one would use triangles, the simplest available pieces of plane area bounded by straight-line segments. The curved surface could then be triangulated, by laying out a set of points on the surface and joining them to neighboring points by straight lines. A spherical surface, for instance, could be approached by a sequence of inscribed polyhedra (not regular), such that all vertices of every polyhedron lay on the surface of the sphere. As the lengths of each individual side of each face of the polyhedron became smaller ("approached zero"), it was believed that the total area of the triangular faces would always have to approach the desired area of the curved surface (Figure 58). The necessary conditions were only that (1) the vertices of all triangles must always lie on the surface; (2) the number of triangles shall increase indefinitely; (3) the lengths of all sides of all triangles shall decrease indefinitely. But a definition, to have any value, must work *all* of the time, not only some of the time. H. A. Schwarz (1843–1921) produced a disconcerting example in which the triangles behaved in accordance with all of the three restrictions, and yet the surface area of the inscribed polyhedron spectacularly failed to converge to the curved surface area. What is particularly im-

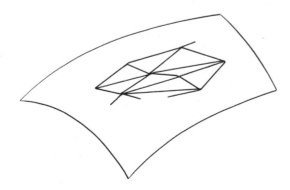

[58]

pressive about Schwarz's example is that the curved surface is nothing fancy or crinkly, but one of the simplest and most innocent of surfaces, the lateral part of a right-circular cylinder. In fact this surface is developable from a plane, so that its area is known in advance.

We describe Schwarz's polyhedron. On the surface of a right-circular cylinder of height h and radius r, select $2n$ vertical lines, equally spaced, and $2n^3$ horizontal equally spaced circles. Join alternate intersection points with straight-line segments as indicated in Figure 59. There are $4n^4$ triangles forming an "accordion-like" polyhedron inscribed in the cylinder. Now let n become indefinitely large. All sides of all triangles tend toward zero, and certainly the vertices remain on the surface. But the planes of the triangles, like a closing accordion, turn edgewise to the surface instead of approaching it tangentially; and the total surface area of the triangle-faced inscribed polyhedron becomes as large as you please, and

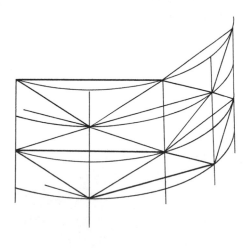

tends to infinity instead of to $2\pi rh$. For a clear
exposition of the mathematics involved, see
p. 350, *Calculus*, John F. Randolph, Macmillan,
1955. (In all references, we shall systematically
list the initial page number only.)

7. The Banach-Tarski theorem was published in *Fundamenta Mathematicae*, Vol. 6 (1924),
p. 244. When one tries to comprehend it, a first
step in the reorganization of preconceived
notions is the concept that a point-set can be
congruent to a proper subset of itself. We offer
a simple example to show that such a set is
possible.

Let P_0 be a fixed point on the unit circle, and
$P_1, P_2, \ldots$ be points on the circle numbered
counterclockwise such that the arc-length between P_n and P_{n+1} is 1; that is, successive P_i
are one radian apart (Figure 60). Let A_0 be the
set of all the P_i. It is an infinite set, because 1 is
incommensurable with 2π. Let φ be the opera-
tion of rotating in the counterclockwise direction through 1 radian around 0, and let $A_1 =
\varphi(A_0)$. Now A_1 is an exact replica of A_0, being

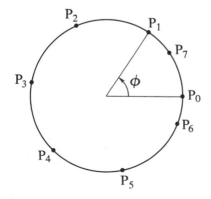

simply a rigid motion rotation of A_0; yet no point of A_1 occupies the position of P_0, because of the irrationality of 2π. In fact A_1 *is* A_0 with one point deleted. Thus A_0 is congruent to a proper subset of A_0. Indeed, by subjecting A_0 to the transformation $n\varphi$, we can obtain A_n, a congruent proper subset "smaller than" A_0 by n points.

This construction is from Hadwiger, Debrunner, and Klee, *Combinatorial Geometry in the Plane,* Holt, Rinehart & Winston, 1964, p. 25.

7. Raphael Robinson, *Fundamenta Mathematicae,* Vol. 34 (1947), p. 246. See also *Amer. Math. Monthly,* Vol 55 (1948), p. 459.

8. "The platonic solids," *Amer. Math. Monthly,* Vol. 76 (1969), p. 192.

11. Tournament: "Mathematics as a creative art," Paul R. Halmos, *Amer. Scientist,* Vol. 56 (1968), p. 375.

15. Figure 61 suggests that all the parallel lines intersect the 45° line at the right kind of points.

18. The proof that the general angle cannot be trisected with ruler and compass is completely developed in many places. See for instance *Ruler and Compasses,* by Hilda P. Hudson, p. 21. This is a Chelsea reprint (1953), bound together with *Squaring the Circle.* Also Felix Klein's *Famous Problems of Elementary Geometry,* Chelsea (1955), p. 5. Courant and Robbins, in *What is Mathematics?* (Oxford Uni-

versity Press, 1941), give a somewhat more sophisticated treatment starting on p. 120.

19. The rectangle problem was suggested by H. D. and S. K. Stein, "On dividing an object efficiently," *Amer. Math. Monthly,* Vol. 63 (1956), p. 111. Another example, now classic, is the search for the solution to the problem of least area in which to reverse a line segment in the plane. It was described in the author's *Excursions in Geometry,* Oxford University Press (1969), p. 147. Since then most of the remaining unsolved part of this famous problem has been conquered by Frederic Cunningham, Jr.: "The Kakeya problem for simply

[61]

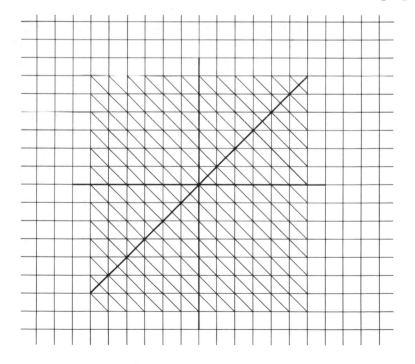

connected and for star-shaped sets," *Amer. Math. Monthly,* Vol. 78 (1971), p. 114.

20. *Gödel's Proof,* Ernest Nagel and James R. Newman, New York University Press, 1958. Alonzo Church and A. M. Turing (separately) took the next logical step by showing that it is impossible to prescribe any systematic procedure for ascertaining whether a given proposition is Gödel-undecidable or not. Church, "An unsolvable problem of elementary number theory," *Amer. Journal Math.,* Vol. 58 (1936), p. 345. Turing, "On computable numbers," *Proc. London Math. Soc.,* (2) Vol. 42 (1937), p. 230.

24. The term "scattering," used in the first edition, has been changed to "dispersal" so that our nomenclature will correspond with that normally used in the current literature.

The deployment problem for $n = 5$ was solved for the plane circular disk by Eric. H. Neville, *Proc. London Math Soc.* (1914), p. 308; but he says nothing about $n > 5$. Inasmuch as the solution for $n = 5$ is not a regular pentagonal arrangement, each value of n probably has to be investigated separately.

The value of k in the deployment problem on the disk does not change with every change in n. We encounter a surprise at the very first stage: $k = r$, the radius of the disk, whether $n = 1$ or $n = 2$. In other words, if we are trying to minimize the greatest distance from any point to the nearest defense unit in a circular country, one unit can do the job precisely as well as two.

At the end of a paper entitled "Unique arrangements of points on a sphere," L. L. Whyte gave a list of (then) unsolved problems associated with deployment and dispersal; *Amer. Math. Monthly,* Vol. 59 (1952), p. 606.

For a summary of work done on the deployment problem by machine approximation for values of *n* up to 10, see "Black box maximization of circle coverage," C. T. Zahn, Jr., *Math. Reviews,* Vol. 29 (1965), p. 316, No. 1583.

See also "Axially symmetric packing of equal circles on a sphere," Michael Goldberg, *Annales Univ. Scient. Budapestiensis, Sectio Math.,* Vol. 10 (1967), p. 39; and "Lower bounds for the disk packing constant," David W. Boyd, *Math. Comp.,* Vol. 24 (1970), p. 697.

26. The two central spheres: The standard lattice packing gives a total of 18 spheres. The question is, can there be more? Raphael M. Robinson, *Notices, Amer. Math. Soc.,* Vol. 16 (1969), p. 422.

27. Hugo Steinhaus, *Mathematical Snapshots,* third American edition, Oxford University Press, 1969, p. 300.

29. C. S. Ogilvy, "An interception problem," *Journal of the Institute of Navigation,* Vol. 5 (1956), p. 89.

31. The problem is closely allied with certain voting matrices. See Manfred Kochen, "A mathematical formulation of influence distri-

butions in decision-making groups," *Journal Soc. Industrial and Applied Math.,* Vol. 6 (1958), p. 199.

34. Freund, "Round robin mathematics," *Amer. Math. Monthly,* Vol. 63 (1956), p. 112. The bridge problem is older: R. E. Moritz, *Amer. Math. Monthly,* Vol. 38 (1931), p. 340.

For a summary of some of the known theory on ranking, see *Topics on Tournaments,* John W. Moon, Holt, Rinehart & Winston, 1968.

In 1963 John W. Ward of Stouffville, Ontario, wrote me concerning bowling tournament schedules developed by him and later acquired by the American Bowling Congress. These schedules incorporate additional "mix" conditions. Mr. Ward was then "working on a textbook explaining how it is all done," but he was eighty-one at the time, and I have not heard whether he completed the work.

34. The quotation on the Traveling Salesman problem is from the discussion in *The New World of Math,* George A. W. Boehm, Dial Press, 1959. A solution for the U.S. problem (slightly modified) appears on p. 116.

See "The Highway inspector and the salesman," Ch. 9 of Sherman K. Stein's *Mathematics, The Man-made Universe,* W. H. Freeman & Co., 1969. Some advances toward a general solution were made by Kenneth Lebensold in 1967; *Amer. Math. Monthly,* Vol. 74, p. 552.

35. Franz Hohn, "The mathematical aspects of switching," *Amer. Math. Monthly,* Vol. 62 (1955), p. 75. For an elementary treatment, see

Notes

Boolean Algebra, S. A. Adelfio and C. F. Nolan, Hayden, N.Y., 1964.

37. See, for instance, *The Three Body Problem in Nuclear and Particle Physics,* J. S. C. McKee and P. M. Rolph, Elsevier Publishing Co., 1970.

38. "College admissions and the stability of marriage," D. Gale and L. S. Shapley, *Amer. Math. Monthly,* Vol. 69 (1962), p. 9.

39. At the beginning of this chapter in the first edition there was a bridge problem of Ulam, who asked for a deal such that (1) North and South can make, against any defense, a grand slam in *any* suit provided that suit is trump, but (2) against good defense they can make only five no-trump. Three solutions were soon forthcoming, two from amateurs and one from the professional mathematician Marion K. Fort. Fort's was the simplest:

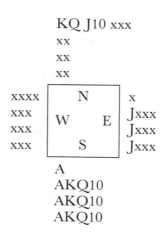

The point is of course that anyone picking up the South hand would have difficulty refraining from ending up in seven no-trump—the wrong contract.

40. Martin Gardner, whose column of mathematical games appears monthly in *Scientific American,* gave generous assistance when I was selecting problems for this edition. He calls the paradox "The Unexpected Hanging," and the discussion in Chapter 1 of his book by that name is the best of many that have appeared in recent years. Martin Gardner, *The Unexpected Hanging and Other Mathematical Diversions,* Simon & Schuster, 1969.

42. Leonard Euler (1707–83) said to be the most prolific mathematician in history.

E. T. Parker, R. C. Bose, and S. S. Shrikhande solved the tenth-order problem, thus disproving Euler's conjecture. Their discovery made the front page of the *New York Times* (April 26, 1959), a most unusual occurrence for any topic in mathematics. Gardner's column in the November 1959 *Scientific American* gave the full story, and the front cover of the magazine carried a color picture of the 10×10 Graeco-Latin square of our Figure 12.

Note in conection with the subsquare of order 3: Interchanging the order (position) of any two rows or of any two columns of a Graeco-Latin square results in another Graeco-Latin square. But the people who work with these things consider that this is such a trivial change that they call two such squares the same square. This means that one cannot simply shift a column out of the subsquare to break it up: that doesn't count!

47. The theoretical machine used in Rado's problem of the three cards is a so-called *Turing machine,* after its inventor, A. M. Turing. The program that produces ten 1's is:

Card 1		Card 2		Card 3	
B	0-R-2	B	B-R-3	B	1-L-3
0	1-R-1	0	0-R-2	0	1-R-2
1	1-R-STOP	1	B-R-2	1	0-L-1

47. The problem of the balls is equivalent to a ranking problem described by Ford and Johnson, *Amer. Math. Monthly,* Vol. 66 (1959), p. 387. Steinhaus gives an incomplete discussion, *Mathematical Snapshots*, third edition, p. 56.

48. The definition of the theory of games is quoted directly from *The Compleat Strategyst,* by J. D. Williams, McGraw-Hill, 1954, p. 215. The book is written entirely without mathematics, and is a noble effort to present a technical subject non-technically. There are times when it seems that the price is rather high: some passages suffer from a thirst that could quickly be quenched by a short draught of mathematics. But on the whole, the book succeeds in its aim. At the opposite extreme, *The Theory of Games and Economic Behavior,* by John

von Neumann and Oskar Morgenstern, Princeton University Press, 1944, is rough going.

I cannot resist the temptation to tell a famous anecdote about von Neumann, who was one of the great intellects at the Institute for Advanced Study until his untimely death in 1957. A friend of his, out strolling with him one day, gave him the problem of the Industrious Bee (see the author's *Mathescope,* p. 39), which has a very simple solution if you happen to attack it from the proper angle. After walking on for a few steps, von Neumann turned to his companion and gave him the correct answer. His friend smiled and remarked "You saw the trick because you are a mathematician. Most people try to solve it by summing convergent infinite series, which is quite a project." "I know," replied von Neumann, dryly, "that's what I just now did."

51. The tetrahedron problem: No. 4516, Victor Thebault, *Amer. Math. Monthly,* Vol. 59 (1952), p. 702.

52. The parabola problem: No. 4241, R. Goormaghtigh, *Amer. Math. Monthly,* Vol. 54 (1947), p. 168.

53. The solution to the Lebesgue covering problem is known to have area A somewhere within the following bounds, for unit diameter:

$$.8257 \leq A \leq .8441$$

The lower bound is equal to $\pi/8 + \sqrt{3}/4$. The upper bound is the area of Sprague's solution of Figure 15. The problem appears on p. 100 of

I. M. Yaglom and V. G. Boltyanski, *Convex Figures,* Holt, Rinehart & Winston, 1961, which contains many other unsolved problems. See also "Borsuk's problem and related questions," Branko Grünbaum, *Convexity; Proc. Symposia Pure Math.,* Amer. Math. Soc., Vol. 7 (1963), p. 274. Sprague's result is in "Uber ein elementares Variationsproblem," *Matematisk Tidsskrift,* 1936, p. 96. An excellent summary of recent results, with statements of many more unsolved problems, is to be found in "Intersection and covering problems in convex sets," G. D. Chakerian, *Amer. Math. Monthly,* Vol. 76 (1969), p. 753.

The Lebesgue problem in three dimensions: Herbert Meschkowski, *Unsolved and Unsolvable Problems in Geometry,* Unger, 1966, p. 70.

55. The theorem on quadrilaterals is given by G. D. Chakerian and L. H. Lange in "Geometric extremum problems," *Math. Magazine,* Vol. 44 (1971), p. 57.

56. Figure of minimum area with given circumference and diameter: M. Scholander, "On certain minimum problems in the theory of convex curves," *Trans. Amer. Math. Soc.,* Vol. 73 (1952), p. 139.

56. The two questions about the perpendicular chords are posed by Nicholas D. Kazarinoff in a book called *Analytic Inequalities,* Holt, Rinehart & Winston, 1961, p. 85. The first conjecture is due to Peter Ungar, of New York University.

56. Equichordal curves: G. Dirac, *Journal*

London Math. Soc., Vol. 27 (1952), p. 429.

The last problem on page 56 of the first edition has been solved in the negative by Hallard T. Croft, "Two problems on convex bodies," *Proc. Cambridge Phil. Soc.*, Vol. 58, Part 1 (1962), p. 1.

57. Murray Klamkin, *Math. Magazine,* Vol. 42 (1969), p. 223.

57. Stanislav M. Ulam, at the Los Alamos Scientific Laboratories in New Mexico, has written No. 8 of the Interscience Tracts in Pure and Applied Mathematics, called *A Collection of Mathematical Problems,* Interscience, 1960. Several of our problems come from this book, which we shall hereafter refer to as "Ulam." The equilibrium problem is mentioned on page 38.

58. HA-measurements: *Amer. Math. Monthly,* Vol. 76 (1969), p. 539.

59. Goldberg on rotors: *Amer. Math. Monthly,* Vol. 64 (1957), p. 76. Also *Mathematics of Computation,* Vol. 14 (1960), p. 235, where a list of thirty-six pertinent references is given. He has solved the problem of least area stated near the bottom of p. 58 of the first edition; *Math. Reviews,* Vol. 33 (1969), p. 1164.

59. "Is every polygonal region illuminable from some point?" Victor Klee, *Amer. Math. Monthly,* Vol. 76 (1969), p. 180.

61. Lattice covering: Waclaw Sierpinski, *A Selection of Problems in the Theory of Numbers,* Pergamon, 1964, p. 16.

Notes

62. Triangle: Hugo Steinhaus, *One Hundred Problems in Elementary Mathematics*, Basic Books, 1964, p. 48.

The tetrahedron problem was communicated to me by Robert Spira of Michigan State University.

63. Squares in curves. First question: *Proc. Symposia in Pure Math.*, Amer. Math. Soc., Vol. 7 (1963), p. 19. Second question: An everywhere differentiable curve has no corners or cusps. For such a curve the answer to this question is known to be yes. "Inscribed squares in plane curves," R. P. Jerrard, *Trans. Amer. Math. Soc.*, Vol. 98 (1961), p. 234. An allied problem has been solved: It is possible to place a square table on a wavy floor so that all legs touch the floor and the table-top is horizontal; but there exist wavy floors for which this cannot be done with a regular pentagonal table. Roger Fenn, "The table theorem," *Bull. London Math. Soc.*, Vol. 2 (1970), p. 73. See also "Figures inscribed in convex sets," H. E. Eggleston, *Amer. Math. Monthly*, Vol. 65 (1958), p. 76.

63. Equilateral triangles: Murray Klamkin, Problem 5465, *Amer. Math. Monthly*, Vol. 74 (1967), p. 207.

64. The question on the crated sphere is asked by H. S. M. Coxeter on p. 495 of *Convexity* (see note to p. 53 above). See also G. C. Shephard, "A sphere in a crate," *Jour. London Math. Soc.*, Vol. 40 (1965), p. 433.

64. Inscribed m-gon: The case $n = 4$ appears

to have been solved. If m is an integral multiple of 4, the obvious solution is also the correct one. If m is not an integral multiple of 4, the inscribed m-gon "has an orientation half way between two successive orientations in which a side of the m-gon is parallel to a side of the square." Frederick J. Fuglister and Clifford M. Bryant, Jr., Colby College (unpublished communication, 1971).

64. Three concurrent chords at 60°. Steinhaus, *Polish Academy of Sciences,* Class 3, 1957, p. 595. Also *Mathematical Snapshots,* p. 153 of the third edition.

64. Non-bendability of a convex surface: Hilbert and Cohn-Vossen, *Geometry and the Imagination,* Chelsea, 1952, p. 230. Re-entrant polyhedra, *ibid.,* p. 290.

65. Six conics: *Amer. Math. Monthly,* Vol. 50 (1943), p. 260. "Rational distances on a conic section," D. E. Daykin, *Amer. Math. Monthly,* Vol. 77 (1970), p. 314.

65. The question on skew lines in 3-space was suggested by Littlewood. See also *Mathematical Puzzles and Diversions,* by Martin Gardner, Crowell 1961, p. 105.

The four problems appearing on p. 62 of the first edition have all been solved. The answer to (1) is yes. "On the Steinhaus billiard table problem," H. T. Croft and H. P. F. Swinnerton-Dyer, *Proc. Cambridge Phil. Soc.,* Vol. 59 (1963), p. 37. (2) Also yes. "Mathematical games," Martin Gardner, *Scientific American,* September 1963, p. 248.

"Can the inscribed triangle ever have the least perimeter?" No. *Amer. Math. Monthly,* Vol. 69 (1962), p. 672, No. 4964.

"Is the maximum plane section of a tetrahedron its largest face?" Yes. *Amer. Math. Monthly,* Vol. 70 (1963), p. 1108, No. 5006. See also Vol. 75 (1968), p. 34, for the extended problem in *n* dimensions, where the answer is not always the same.

66. Maximum section of a right-circular cylinder, *Amer. Math. Monthly,* Vol. 60 (1953), p. 715.

The answer to the question on p. 65 of the first edition is no. Free Jamison, "An impossible dissection," *Amer. Math. Monthly,* Vol. 69 (1962), p. 550. The dissection of Figure 20 on the same page has been shown by P. J. Federico to be essentially unique.

67. "Can every triangle be divided into *n* triangles similar to it?", *Amer. Math. Monthly,* Vol. 77 (1970), p. 867.

67. The dissection into twenty-four different squares, with a fascinating account of its discovery, is described in November 1958 *Scientific American,* p. 142. The problem has a direct connection with the flow of current in electrical networks, consideration of which led to its ultimate solution.

68. Dissection of the 2×1 rectangle: P. J. Federico, *Jour. of Combinatorial Theory,* Vol. 8 (1970), p. 244. See also the paper that precedes it by R. L. Brooks.

69. For further problems on sums of consecu-

tive squares, both solved and unsolved, see the paper by H. L. Adler and Bro. U. A. Alfred, *Amer. Math. Monthly,* Vol. 71 (1964), p. 749.

70. Mrs. Perkins' Quilt: $S(n)$ is given for $n \leqslant 12$ in Martin Gardner's column in *Scientific American,* September 1966, p. 264.

71. Dissection of tetrahedra: "Two more tetrahedra equivalent to cubes by dissection," Michael Goldberg, *Elemente der Mathematik,* Vol. 24 (1969), p. 130.

72. R. C. Read has determined the number of polyominoes of orders 8, 9, and 10: "Contributions to the cell growth problem," *Canadian Jour. Math.,* Vol. 14 (1962), p. 1. Read's methods seem not to lead in the direction of a general formula for order n.

David A. Klarner discusses the dissection of rectangles into congruent polyominoes, and notes several unsolved problems in this area. *Jour. Combinatorial Theory,* Vol. 7 (1969), p. 107. See also Solomon W. Golomb, *Polyominoes,* Charles Scribner & Sons, 1965.

72. R. B. Kershner gives an excellent summary of known information on the general tiling problem, "On paving the plane," *Amer. Math. Monthly,* Vol. 75 (1968), p. 839. Rep-tiles: Martin Gardner, *The Unexpected Hanging,* Ch. 19. Integer triangle tiling: D. C. Kay gives a formula for doing the rational tiling, *Amer. Math. Monthly,* Vol. 73 (1966), p. 903.

74. "The packing of equal spheres," C. A. Rogers, *Proc. London Math. Soc.,* Series 3, Vol. 8

Notes

(1958), p. 609. Also H. S. M. Coxeter, *Introduction to Geometry,* Wiley, 1961, p. 457.

Twelve Pennies: *Math. Reviews,* Vol. 34 (1970), p. 1164, No. 6380.

75. Packing of squares: "Some packing and covering theorems," J. W. Moon and Leo Moser, *Colloquium Mathematicum,* Vol. 17 (1967), p. 103. They show that $\% < T \leqslant 2$.

Packing of dissected disks: *Amer. Math. Monthly,* Vol. 75 (1968), p. 195.

76. "Stability of polyhedra," Richard K. Guy, *Soc. Indust. & Appl. Math. Review,* Vol. 11 (1969), p. 78.

79. "Paths of minimal length within a hypercube," Richard A. Jacobson, *Amer. Math. Monthly,* Vol. 73 (1966), p. 868.

Convex Polytopes, Branko Grünbaum, Interscience (Wiley), 1967.

80. "Polytopes, graphs and complexes," Branko Grünbaum, *Bull. Amer. Math. Soc.,* Vol. 76 (1970), p. 1131.

80. Equilateral triangles: J. Gallego-Diaz, *Amer. Math. Monthly,* Vol. 60 (1953), p. 336. Quadrilaterals: Josef Langr, *ibid.,* p. 551. The ratio of similitude is the hard part. It seems not too difficult to show that the quadrilaterals are similar; any quadrilateral can be divided into two triangles by a diagonal, and the problem thus reduced to one of similarity of the triangles. John A. Wright, of Chester Basin, Nova Scotia, has sent me that much of the solution.

83. The quotations are from "Thebault—

the number theorist," by E. P. Starke, part of a tribute published in the October 1947 *Amer. Math. Monthly,* Vol. 54, p. 443. The quoted problem is No. 3886, p. 482 of Vol. 45.

86. Pierre de Fermat (1601–65). The 5500 figure was announced in "A proof of Fermat's last theorem for all prime exponents less than 5500," *Soviet Math. Doklady* (translated), Vol. 11 (1970), p. 188.

90. For a tabulation of the known data on F_n, see "A report on primes of the form $k \cdot 2^n + 1$ and on factors of Fermat numbers," Raphael M. Robinson, *Proc. Amer. Math. Soc.,* Vol. 9 (1958), p. 673; also "New factors of Fermat numbers," Claude P. Wrathall, *Math. Comp.,* Vol. 18 (1964), p. 324.

91. F_7 and F_6: J. C. Morehead, "Note on Fermat's numbers," *Bull. Amer. Math. Soc.,* Vol. 11 (1905), p. 543; J. C. Morehead and A. E. Western, *ibid.,* 16 (1910), p. 1. The factorization of F_7 is given by Michael A. Morrison and John Brillhart in *Bull. Amer. Math. Soc.,* Vol. 77 (1971), p. 264. When congratulated on their feat, Brillhart remarked, "We are indeed pleased to have vanquished this monster."

93. "Groups, graphs, and Fermat's last theorem," Steven Bryant, *Amer. Math. Monthly,* Vol. 74 (1967), p. 152.

95. The proof of the prime number property of Pascal's Triangle is given in the author's *Through the Mathescope,* p. 137.

96. Coefficients divisible by composite n:

Notes

E 1154, *Amer. Math. Monthly,* Vol. 61 (1954), p. 712.

"How often does an integer occur as a binomial coefficient?", David Singmaster, *Amer. Math. Monthly,* Vol. 78 (1971), p. 385.

97. Successive differences: "On a conjecture concerning the primes," R. B. Killgrove and K. E. Ralston, *Mathematical Tables and Aids to Computation,* Vol. 13 (1959), p. 121. This is the publication whose name was changed to *Mathematics of Computation* in 1960.

98. "Some conjectures associated with the Goldbach conjecture." Under this title, I. A. Barnett and Ted Cook of the University of Cincinnati, wrote:

"The first conjecture, a stronger form of the Goldbach conjecture for odd numbers, says that every odd number $2k - 1 = x + 2y$, where x and y are primes, $k \geqslant 4$. The second conjecture is that it is possible to find a representation of every odd number of the form $6k + 1$ or $6k + 5$ as $2x' + 3y'$ (x', y' prime) $k \geqslant 3$, for which either the x' or the y' appears as one of the primes in *some* representation of $2k - 1$ as the sum of a prime plus the double of a prime. Both conjectures have been verified to about 15000." *Amer. Math. Monthly,* Vol. 68 (1961), p. 711.

99. The quotations are from Ulam, p. 120. The second Ulam sequence was communicated to the author by letter. See also "An observation on the distribution of the primes," M. Stein and

S. M. Ulam, *Amer. Math. Monthly,* Vol. 74 (1967), p. 43.

101. Palindromic pairs of primes: I. A. Barnett, *Amer. Math. Monthly,* Vol. 70 (1963), p. 926, No. 7.

"Palindromes by addition," C. W. Trigg, *Math. Magazine,* Vol. 40 (1967), p. 26. Steven Hollingworth, an undergraduate at Michigan State University, ran the program on the number 196.

105. "A conjecture on consecutive composite numbers," C. A. Grimm, *Amer. Math. Monthly,* Vol. 76 (1969), p. 1126.

Prime primes: The complete list is given in *Math. Magazine,* Vol. 42 (1969), p. 232.

106. For further discussion of the decimilization of ¹⁄₇, see Ch. 2 of the author's *Through the Mathescope.*

107. The present guess is that there are no more prime strings of 1's beyond $n = 23$, but this conjecture is weak, without a shred of supporting mathematical evidence. The only empirical evidence is that computer programs have found that all such numbers are composite for $23 < n \leqslant 359$. "Some factorizations of $2^n + 1$ and related results," John Brillhart and J. L. Selfridge, *Math. Comp.,* Vol. 21 (1967), p. 87 (the pertinent statement is at the top of p. 91). The connection between the two problems is fully explained in Ch. 5 of *Excursions in Number Theory,* C. S. Ogilvy and J. T. Anderson, Oxford University Press, 1966.

Notes

108. n! + 1: *Amer. Math. Monthly,* Vol. 58 (1951), p. 193.

109. W. J. Ellison has an excellent expository paper, "Waring's problem," in *Amer. Math. Monthly,* Vol. 78 (1971), p. 10, with an extensive list of references. The problem is also treated in Rademacher and Toeplitz, *Enjoyment of Mathematics,* Princeton University Press, 1957, Ch. 9. The fifteen integers requiring 8 cubes for their decomposition are 15, 22, 50, 114, 167, 175, 186, 212, 213, 238, 303, 364, 420, 428, and 454. The largest known to require 7 is $8042 = 19^3 + 10^3 + 4^3 + 4^3 + 3^3 + 3^3 + 1^3$.

For the Mordell remark, see *Reflections of a Mathematician,* J. L. Mordell, Canadian Math. Congress, 1959, p. 19. It is known as the "easier" Waring problem.

110. Perfect numbers: Rademacher and Toeplitz, Ch. 19.

The solution of the problem on abundant numbers that appeared on page 93 of the first edition is given in Ogilvy and Anderson, p. 23.

112. Amicable and sociable numbers. In 1939 the score stood at 41 pairs of even amicable numbers and 27 pairs of odd ones for a total of only 68 known pairs. B. H. Brown, "A new pair of amicable numbers," *Amer. Math. Monthly,* Vol. 46 (1939), p. 345. The computer age had not yet begun. By 1970 this number had risen to more than 1000.

Some conditions on the existence of pairs of opposite parity are given by A. A. Gioia,

"Amicable numbers of opposite parity," *Amer. Math. Monthly,* Vol. 74 (1967), p. 969. See also Peter Hagis, Jr., "Relatively prime amicable numbers of opposite parity," *Math. Magazine,* Vol. 43 (1970), p. 14. Hagis has found that if two such numbers exist, each exceeds 10^{60}. *Math. Comp.,* Vol. 24 (1970), p. 963.

The sociable chain of 28 links has smallest member 14316. The 5-link chain was announced in the same paper: P. Poulet, *L'Intermediaire des mathematiques,* Vol. 25 (1918), p. 100.

4-link chains: "On amicable and sociable numbers," Henri Cohen, *Math. Comp.,* Vol. 24 (1970), p. 423. See also the paper immediately following it, and on page 493 of the same volume the review by Daniel Shanks of Elvin J. Lee's *The Discovery of Amicable Numbers.*

114. Double decompositions: If any solution exists for $a^5 + b^5 = c^5 + d^5$, the sum of the two fifth powers exceeds 2.8×10^{14}. This and much other information is collected in "A survey of equal sums of like powers," Lander, Parkin, and Selfridge, *Math. Comp.* Vol. 21 (1967), p. 446. See also "Some new results on equal sums of like powers," Simcha Brudno, *Math. Comp.* Vol. 23 (1969), p. 877, which ends with the statement of another unsolved problem. In this and similar papers, square brackets [ ] indicate the "greatest integer" function.

The Dickson remark was verified by J. Leech, *Proc. Cambridge Phil. Soc.,* Vol. 53 (1957), p. 779.

115. The first problem on consecutive powers is from *Amer. Math. Monthly,* Vol. 39 (1932), p. 175. See also Dickson, Vol. 2, p. 585. The extended problem is from *Amer. Math. Monthly,* Vol. 45 (1938), p. 253.

116. *Amer. Math. Monthly,* Vol. 76 (1969), p. 952, No. 5634.

In the first edition a similar problem was posed: Does the equation $m(m + 1)(m + 2) = n(n + 1)(2n + 1)$ have any solution in positive integers besides the trivial $(m,n) = (1,1)$? It was answered in the negative by Raphael Finkelstein, *Amer. Math. Monthly,* Vol. 73 (1966), p. 471. The proof takes six pages.

The numbers 1, 3, 6, 10, . . . have been called *triangular* numbers, by analogy to the square numbers (see Figure 62). The formula for the sum of the first m triangular numbers is $m(m + 1)(m + 2)/6$. If cannon balls are piled in a triangular pyramid the layers will be triangular numbers; if piled in a square pyramid they will be square numbers. It has been pointed out (by Victor Meally, of Dublin, Ireland) that the negative solution of this problem means that no triangular pyramidal pile can ever contain the same number of cannon

[62]

1   3   6   10

balls that form some square pyramidal pile.

116. The set of fourteen Diophantine problems came from the following sources:

2, 3, 4, 5, 8, and 9(b) are from a paper by Waclaw Sierpinski, "On some unsolved problems of arithmetics," *Scripta Mathematica,* Vol. 25 (1960), p. 125. Sierpinski lists many more in that paper and in his book, *A Selection of Problems in the Theory of Numbers,* Pergamon, 1964.

9(a), 9(c): "Representation of an integer as a sum of four integer cubes and related problems," D. A. Klarner, *Amer. Math. Monthly,* Vol. 74 (1967), p. 531.

6, 7, and 10 are posed by Mordell, *Jour. Lond. Math. Soc.,* Vol. 28 (1953), p. 500.

1, 11, 12, and 13 were given to me by I. A. Barnett, except for the question about asymptotic density.

14 is Mordell's, *Jour. London Math. Soc.,* Vol. 36 (1961), p. 355.

Regarding (4), (*cf.* first edition); it is now known that three consecutive numbers of this form cannot exist. Seppo Hyyro, *Math. Reviews,* Vol. 28 (1964), No. 62.

119. First problem on cubics: Mordell, *Reflections of a Mathematician,* p. 32. Second problem on cubics: H. Davenport, *Proc. Royal Soc.,* Series A, Vol. 272 (1963), p. 285.

120. First "Pythagorean" question: *Amer. Math. Monthly,* Vol. 62 (1955), p. 494; but it has been knocking around for a long time, before and since. In *Amer. Math. Monthly,* Vol. 70 (1963), p. 210, is given the minimum solution

Notes

for all elements *except* the main diagonal an integer. Second "Pythagorean" question: Offered by Solomon W. Golomb in 1971.

120. Walker's problem: Nos. 4326 and 4327, *Amer. Math. Monthly,* Vol. 56 (1949), p. 39, and for proof of the hint, *ibid.,* Vol. 57 (1950), p. 350.

121. "Powerful numbers," Solomon W. Golomb, *Amer. Math. Monthly,* Vol. 77 (1970), p. 848.

123. The table is on page 116 of *Continued Fractions,* C. D. Olds, Random House, 1963. Professor Olds in some further (unpublished) work found that if there is a bound on L it is large $\sqrt{1000099}$ has $L = 2174$. See also Ch. 10 of Ogilvy and Anderson.

124. "Fibonacci representations I and II", L. Carlitz, *Fibonacci Quarterly,* Vol. 6 (1968), p. 193, and Vol. 8 (1970), p. 113. Several other questions are posed at the end of II.

The periodicity problem is stated in two places in *Amer. Math. Monthly,* Vol. 72 (1965): "On second order recurrences," Oswald Wyler, p. 505; "On periodicity in generalized Fibonacci sequences," D. M. Bloom, p. 861. The quotation is his closing remark.

Anyone who intends to investigate the many intriguing properties of this famous sequence would do well to consult the large body of information assembled in the *Fibonacci Quarterly,* Volume 9 of which appeared in 1971. It is the journal of The Fibonacci Association, St. Mary's College P.O., Calif. 94575.

The conjecture that 144 is the only Fibonacci

square number besides 1 (see first edition) has been proved. *Amer. Math. Monthly,* Vol. 71 (1964), p. 221, and by another method in *Fibonacci Quarterly,* Vol. 2 (1964), p. 109. Abstract No. 663-643, *Notices Amer. Math. Soc.,* Vol. 16 (1969), p. 275, states that 1 and 8 are the only Fibonacci *cube* numbers.

128. "Is such a grid possible?" No.

129. That four colors are sufficient to color a map of thirty-four or fewer countries was proved by Philip Franklin of M.I.T., *Journal of Mathematics and Physics,* Vol. 16 (1937), p. 172.

Steinhaus, *Mathematical Snapshots,* has a clearly drawn diagram of a torus map requiring seven colors on p. 274 of the third edition.

133. Ulam, p. 50.

The quotation is from "The elusive fixed point property," R. H. Bing, *Amer. Math. Monthly,* Vol. 76 (1969), p. 119.

134. Enlacement: Ulam, p. 46. Lines of force: Ulam, p. 108.

Two problems from page 114 of the first edition have been solved. (1) "A problem in geometrical probability," Eric Langford, *Math. Magazine,* Vol. 43 (1970), p. 237. (2) "A type of 'Gambler's Ruin' problem," R. C. Read, *Amer. Math. Monthly,* Vol. 73 (1966), p. 177. The problem is replaced by a new unsolved one at the end of the paper.

139. "The problem of the thrown string," J. L. Synge, *Math. Gazette,* Vol. 54 (1970), p. 250.

140. The postage stamp problem has been

"solved" by computer up to $n = 24$ stamps. Although the numbers of ways of folding are less than $n!$, they are nevertheless disconcertingly large, and no formula has been deduced. "A map-folding problem," W. F. Lunnon, *Math. Comp.*, Vol. 22 (1968), p. 193.

140. Random walk: Gian-Carlo Rota, "Combinatorial analysis," in *The Mathematical Sciences* (COSRIMS Report), M.I.T. Press, 1969, p. 201.

There are 4 possible random walks of length 1. At the end of each of these one can take a second step in any of 4 directions, for a total of 4^2 possible walks of length 2. Likewise there will be 4^3 possible walks of length 3, and in general 4^n walks of length n. The walks asked for in the problem comprise a proper subset of these, and therefore must number "something less than 4^n."

141. The quotation is from Mark Kac, "Probability," *The Mathematical Sciences* (COSRIMS Report), M.I.T. Press, 1969, p. 242.

"What is the expected volume of a simplex whose vertices are chosen at random from a given convex body?" Victor Klee, *Amer. Math. Monthly*, Vol. 76 (1969), p. 286.

142. Abstract combinatorial problem: *Amer. Math. Monthly*, Vol. 51 (1944), p. 534.

143. *The Caliban problem.* The probability of drawing k red balls from a bag containing n balls of which r are red is $p = {_rC_k}/{_nC_k}$. One

sees at once by expanding the C's that $p = \frac{1}{2}$ whenever $n = 2k$ and $r = 2k - 1$:

$$p = \frac{(2k-1)(2k-2)\cdots(2k-k+1)(2k-k)}{2k(2k-1)(2k-2)\cdots(2k-k+1)}$$

$$= \frac{1}{2}$$

For $k = 5$ we have the Caliban answer. The question remains whether it is possible to obtain $\frac{1}{2}$ with less cancellation of the factors. One finds by trial that there are always too many prime factors. We display the information for p near $\frac{1}{2}$ in a table. The number in the left-hand column of each row is the number of factors in the numerator of p which are to cancel completely by occurring also in the denominator.

For the cases marked * we consider general k and ask that all but 2 factors cancel in their entirety, leaving

$$\frac{(n-k)(n-k-1)}{n(n-1)} = \frac{1}{2}$$

solving this quadratic for k yields:

$$k = \frac{2n - 1 \pm \sqrt{2n^2 - 2n + 1}}{2}$$

The discriminant must equal a perfect square, say m^2. Solving that now for n in terms of m, we find

$$n = \frac{1}{2}(1 \pm \sqrt{2m^2 - 1})$$

	$k = 5$	$k = 6$
0	This requires a sequence of $2k$ composite integers in the vicinity of 50. No such sequence exists.	
1	$\dfrac{_{28}C_5}{_{32}C_5} < \dfrac{1}{2} < \dfrac{_{29}C_5}{_{33}C_5}$	$\dfrac{_{43}C_6}{_{48}C_6} < \dfrac{1}{2} < \dfrac{_{44}C_6}{_{49}C_6}$
2	$\dfrac{_{22}C_5}{_{25}C_5} < \dfrac{1}{2} < \dfrac{_{23}C_5}{_{26}C_5}$	$\dfrac{_{35}C_6}{_{39}C_6} < \dfrac{1}{2} < \dfrac{_{36}C_6}{_{40}C_6}$
3	$*$	$\dfrac{_{27}C_6}{_{30}C_6} < \dfrac{1}{2} < \dfrac{_{28}C_6}{_{31}C_6}$
4	The Caliban case: $\dfrac{_{9}C_5}{_{10}C_5} = \dfrac{1}{2}$	$*$
5		The Caliban case: $\dfrac{_{11}C_6}{_{12}C_6} = \dfrac{1}{2}$

The various m which make $2m^2 - 1$ a perfect square are well known to be the denominators of every second convergent of the continued fraction development of $\sqrt{2}$.

The first (usable) m is 5, which says that $n = 4$, $k = 1$, $r = 2$ constitute a solution. It is the rather trivial one of drawing one ball from a bag containing two reds and two others. Clearly $p = \frac{1}{2}$.

The next (usable) m is 29; $n = 21$, $k = 6$, $r = 19$. Thus the probability of drawing six red balls from a bag containing nineteen reds and two others is $\frac{1}{2}$. We note that this is *in addition* to the Caliban case, one of which occurs for every k. Thus the solution of (2) is not unique; but we have also proven that 5 is not a suitable k, allowing the conclusion that the solution of (1) is unique.

We have also located some other k for which the solution is surely not unique, the next one being $k = 35$, for which the pair $n = 120$, $r = 118$ is a solution.

There remains unanswered the question: For what additional k is the solution non-unique because all but 3, or all but 4, or all but ... k factors cancel completely? The general Diophantine equations for these cases are of degree 3, 4, ... k respectively, and are consequently untidy.

143. The problem of the fewest intersections is discussed by Richard K. Guy in *NABLA*, the journal of the Malayan Mathematical Society, Singapore, June 1960, p. 68. There is a further discussion (but no solution) in *Proc. Edinburgh Math. Soc.*, Vol. 13 (II), (1963), p. 333.

144. "On a problem of Paul Erdös," E. Altman, *Amer. Math. Monthly,* Vol. 70 (1963), p. 148.

146. The proofs of the partitioning formula

Notes

are given in Yaglom and Yaglom, *Challenging Mathematical Problems,* Holden-Day, 1964, p. 108. Note that the formula is less by n than ours because the polygon is not inscribed in a circle, and thus the outside set of n regions is omitted.

147. Ulam, No. 5, p. 40.

148. H. D. and S. K. Stein, "On dividing an object efficiently," *Amer. Math. Monthly,* Vol. 63 (1956), p. 111. Quadrilateral: E 2185, *Amer. Math. Monthly,* Vol. 77 (1970), p. 531.

149. Minimal surfaces in a wire cube. Courant and Robbins, *What is Mathematics?,* pp. 387, 397, 502. Also see Fig. 212, p. 549, of D'Arcy Thompson, *On Growth and Form,* Cambridge University Press, 1952.

150. The minimal motion problems are quoted in their entirety from Ulam, p. 79, No. 9.

151. The three problems of Erdös. (1) *Amer. Math. Monthly,* Vol. 56 (1949), p. 343. (2) No. 3835 in the *Dunkel Memorial Problem Book* published by the Mathematical Association of America in 1957. (3) Ulam, p. 27, No. 9.

152. Smallest coefficients: *Amer. Math. Monthly,* Vol. 57 (1950), p. 264.

153. An algebraic number is any number, real or complex, that satisfies some equation of the form

$$a_n x^n + a_{n-1} x^{n-1} + \cdots + a_1 x + a_0 = 0,$$

where the coefficients $a_0, a_1, a_2, \ldots$ are integers, and n is a positive integer. $(\pi + e)$ is discussed by W. S. Brown in "Rational exponential expressions and a conjecture concerning π and e," *Amer. Math. Monthly*, Vol. 76 (1969), p. 28. Some progress has recently been made on the problem by David A. Brubaker, *Math. Magazine*, Vol. 44 (1971), p. 267.

The last problem of Chapter 9 of the first edition was solved by Neil Ashby, *Amer. Math. Monthly*, Vol. 71 (1964), p. 430.

Index

Index

Index

Index

Index

Index